4th International Conference on Energy Materials and Applications (ICEMA 2019)

IOP Conference Series: Materials Science and Engineering
Volume 628

Beijing, China
11 – 13 May 2019

ISBN: 978-1-5108-9832-5
ISSN: 1757-8981

Printed from e-media with permission by:

Curran Associates, Inc.
57 Morehouse Lane
Red Hook, NY 12571

Additional copies of this publication are available from:

Curran Associates, Inc.
57 Morehouse Lane
Red Hook, NY 12571 USA
Phone: 845-758-0400
Fax: 845-758-2633
Email: curran@proceedings.com
Web: www.proceedings.com

Table of contents

Volume 628

2019 4th International Conference on Energy Materials and Applications

11–13 May 2019, Beijing, China

Accepted papers received: 27 August 2019
Published online: 8 October 2019

Preface

Preface

Committees

Peer review statement

Papers

Material Property Analysis

Influence of flame retardants on the mechanical properties of glass fiber reinforced compounds in polymeric matrix epoxy and polyester

C F Pérez-Salinas, C B Castro, E R Valencia and O F Freire.....1

Influence of steps temperature on microstructure and hardness of TiO_2 thin films deposited by co-sputtering

L García González, J B Santaella González, F López Huerta, E. Díaz Trueba, L Zamora Peredo, C Zuñiga Islas and C Guarneros Aguilar.....8

Biocompatibility and surface properties of hydrogenated amorphous silicon-germanium thin films prepared by LF-PECVD

F López Huerta, R M Woo García, L García González, A L Herrera May, W Calleja Arriaga, R Vega and E Soto.....15

Modified $BaTiO_3$/PVDF/Ni composites with high dielectric constant and low dielectric loss

Lin Gao, Zhengguang Sun, Junlong Yao and Feng Lu.....22

Study of nano bioactive glass for use as bone biomaterial comparison with micro bioactive glass behaviour

N Rocton, H Oudadesse, S Mosbahi, L Bunetel, P Pellen-Mussi and B Lefeuvre.....26

Development of a PEMFC dynamic model and the application to the analysis of fuel cell vehicle performance

Wei Liu, Zhijun Peng, Bill Kim, Bo Gao and Yiqiang Pei.....36

Effect of sintering parameters on the densification and hardness of NiAl-CNTs composite

M A Awotunde, O O Ayodele, A O Adegbenjo, A M Okoro, M B Shongwe and P A Olubambi.....42

Tensile properties of the reclaimed steel rail flash butt joint

Anuar Pawan, Zuraidah Salleh, Akid H Zulkefli, Ya'kub Md Taib, Anizahyati Alisibramulisi and Nik Roslin Masdek.....48

Material Science

Kinetic study of the photocatalytic oxidation of ethylene over TiO2 thin films

Rodica-Elisabeta Stroe and Lasse A. Rosendahl.....57

Characterisation and wear behaviour of rolled carbon steel in Dot 4 brake fluid

Olawale O Ajibola, Ojo J Akinribide, Samuel O Akinwamide and Peter A Olubambi.....63

Fabrication, experimental investigation of jute fiber reinforced epoxy composites and hybrid composites

Tazbiul Mahmud Aranno, Sobahan Mia, Md. Shafayet Ullah Khan, M. A. Mehedi Hasan, Nyhan Shams Supto, Abu Shaid Sujon, Adib Bin Rashid and Md. Shafiul Ferdous.....70

Wear behaviour of cast aluminium silicon (Al-Si) alloy in Dot 4 brake fluid

Olawale O Ajibola, Sunday Aribo, Oladeji O Ige, Ojo J Akinribide, Samuel O Akinwamide and Peter A Olubambi.....76

Study of photovoltaic solar materials development trend by deductive logic method using big data

Yi-Huei Su and Jung-Hua Wu.....84

Effect of heat treatment on wear behaviour of rolled carbon steel in DOT4 brake fluid

Olawale O Ajibola, Ojo J Akinribide, Samuel O Akinwamide and Peter A Olubambi.....96

IOP Conf. Series: Materials Science and Engineering **628** (2019) 011001 doi:10.1088/1757-899X/628/1/011001

PREFACE

The 4th International Conference on Energy Materials and Applications (ICEMA 2019) was held in Tsinghua University on May 11-13, 2019. ICEMA2019 is an international conference covering research and development in the field of Energy Materials and Applications, which was co-sponsored by Hong Kong Chemical, Biological & Environmental Engineering Society (HKCBEES) and Department of Electrical Engineering of Tsinghua University. The major goal of this conference is to bring together academic scientists, engineers, industry researchers in order to exchange and share their experiences and research results, and discuss the practical challenges encountered and the solutions adopted.

The proceedings present a selection from papers submitted to the conference from universities, research institutes and industries. 14 papers were accepted from 33 presentations from 9 countries and districts: Australia, China, Denmark, Ecuador, France, French Polynesia, Hong Kong, Indonesia and México, providing research in Thermal and Photovoltaic System, Renewable Energy, New Materials, Power System. All of the papers were subjected to peer-review by conference committee members and international reviewers. For this journal proceeding, the selection of papers was strictly based on the quality and relevance to the conference. Here we would like to thank all the technical program committee members who made great efforts on paper reviewing.

A word of special thanks is given to our plenary speakers and keynote speakers who are pleased to make contributions to our conference and share their new research ideas with us. They are Prof. Qing Wang from Pennsylvania State University; Prof. Qiang Zhang from Tsinghua University; Prof. Jesus Toribio from University of Salamanca, Prof. David T.W. Lin from University of Tainan and Assoc. Prof. Yingying Zhang from Tsinghua University.

In the morning of May 13, besides the plenary speech, there is a Young Scholars' Forum. Five outstanding speakers delivered their presentations around the theme of "Energy Materials and Technology". They are Assoc. Prof. Dr. Xihong Peng from Arizona State University; Assoc. Prof. Chenyi Yi from Tsinghua University; Assoc. Prof. Guangzu Zhang from Huazhong University of Science and Technology; Prof. Rujun Ma from Nankai University and Assoc. Prof. Xiaoshi Qian from Shanghai Jiao Tong University.

Tsinghua University was established in 1911 under the name "Tsing Hua Imperial College," initially as a preparatory school for students who would be sent by the Chinese government to study in the United States. The Department of Electrical Engineering of Tsinghua University, founded in 1932, is one of the three earliest engineering departments established by Tsinghua University. In 1989, the former three majors of power system automation, high voltage technology and electrical engineering were merged into a major of "electrical engineering and automation", which was listed in the national specialty recommendation catalogue. This discipline became the first batch of the key discipline and the doctoral degrees conferring unit of first-level disciplines. Thanks to the support of all the teachers and students of department of Electrical Engineering, we can have this honor to hold this conference in Tsinghua

University successfully. This chance makes the experts and scholars from all over the world know more about the history of Tsinghua University and feel the academic atmosphere, broaden our view and learn a lot from each other.

Last but not least, we'd like to again express our heartfelt appreciation to our chairs, sponsors, technical program committee members, organizing committee members, authors and delegates, who made a lot of efforts and contributions year by year. We would also like to reiterate our gratitude to our host the Department of Electrical Engineering of Tsinghua University, for the wonderful arrangement and support they have provided. Thanks to your support and help, we can hold this conference successfully and always keep making progress.

We are looking forward to seeing you in ICEMA2020!

Conference chair and guest editor,
Assoc. Prof. Qi Li,
Tsinghua University, China

IOP Conf. Series: Materials Science and Engineering **628** (2019) 011001 doi:10.1088/1757-899X/628/1/011001

Statement of Peer Review

All papers published in this volume of IOP Conference Series: Materials Science and Engineering have been peer reviewed through processes administered by the proceedings Editors. Reviews were conducted by expert referees to the professional and scientific standards expected of a proceedings journal published by IOP Publishing.

IOP Conf. Series: Materials Science and Engineering **628** (2019) 011002 doi:10.1088/1757-899X/628/1/011002

COMMITTEES

Advisory Chairs

Prof. Zhang Qiang, Tsinghua University, China
Prof. Qing Wang, Pennsylvania State University, USA

Conference Chairs

Assoc. Prof. Qi Li, Tsinghua University, China
Prof. David Lin, University of Tainan

Program Chairs

Prof. Jesus Toribio, University of Salamanca, Spain
Assoc. Prof. Chenyi Yi, Tsinghua University, China
Prof. Nezihe Ayas, Eskisehir Technical University,Turkey
Prof. Chun-Ping Jen, Chung Cheng University

Technical Committee

Dr. Miguel Lorenzo, University of Salamanca, Spain
Dr. S. Harish, Kyushu University, Japan
Senior Lecturer Dr. Siti Ujila Masuri, Universiti Putra Malaysia, Malaysia
Dr. Dai-Viet N. Vo, Universiti Malaysia Pahang, Malaysia
Dr.Muhammad Najam Khan, BUITEMS, Pakistan
Prof. Khaled Touafek, Unité de Recherche Appliquée en Energies Renouvelables, Algeria
Assoc. Prof. Dr. Sushil Kumar Jain, Manipal University Jaipur, India
Assis. Prof. Dr. Francesca Scalisi, University of Palermo, Italy
Dr. Francisco Lopez-Huerta, Centro de Investigación en Micro y Nanotecnología, Mexico
Prof.Jui Ching Hsieh, Chin-Yi University of Technology
Assoc. Prof.Hung-Pin Hsu, Ming Chi University of Technology
Prof. Macodou Thiam, The University Institute of Technology of the University of THIES, Senegal
Prof.Dr. Najma Laaroussi, Mohammed V University of Rabat, Morocco
Assoc. Prof. Ngoc-Chung Le, Da Lat University, Viet Nam
Dr. Haryati Yaacob, Universiti Teknologi Malaysia, Malaysia
Prof. Jun Peng, University of Bedfordshire, UK
Asst. Prof. Dr.Xiaoying Lu, Technological and Higher Education Institute of Hong Kong, Hong Kong
Prof.Osman Adiguzel, Firat University, Turkey
Prof. Balasubramanian, Anna University India, India
Prof. Dr. Hassane Oudadesse, University of Rennes 1, France
Dr. B. Narayanamoorthy, Jiangsu University, China

IOP Conf. Series: Materials Science and Engineering **628** (2019) 011003 doi:10.1088/1757-899X/628/1/011003

Peer review statement

All papers published in this volume of *IOP Conference Series: Materials Science and Engineering* have been peer reviewed through processes administered by the proceedings Editors. Reviews were conducted by expert referees to the professional and scientific standards expected of a proceedings journal published by IOP Publishing.

IOP Conf. Series: Materials Science and Engineering **628** (2019) 012001 doi:10.1088/1757-899X/628/1/012001

Influence of flame retardants on the mechanical properties of glass fiber reinforced compounds in polymeric matrix epoxy and polyester

C F Pérez-Salinas, C B Castro, E R Valencia, O F Freire

Technical University of Ambato, Faculty of Civil and Mechanical Engineering, Material and manufacturing research group, Av. Los Chasquis and Río Payamino, Ambato, Ecuador.

corresponding author's Email: cf.perez@uta.edu.ec

Abstract. In the present study, polymeric epoxy and polyester resin reinforced with glass fiber compounds were prepared by adding of flame retardants based on aluminum and magnesium hydroxide in order to evaluate their influence on mechanical properties. An experimental DOE design was developed with two qualitative factors: the resin and retardant type; with two levels each and a quantitative factor: volumetric composition of the retardant with three levels (3%, 6% and 9%). With all possible combinations; tensile, flexural and impact tests were carried out. The results showed that the addition of retardants in different percentages has different influence for each mechanical property. It was evidenced that the 9% HA Aluminum Hydroxide has no significant effect on the tensile strength and its modulus compared to Magnesium Hydroxide HM. Conversely; on the flexural strength, flexural modulus and rupture energy the HM has a slightly lower effect with respect to the HA. For flexural strength and its modulus, the best proportion of retardant was 6%. 3% is the recommendation for rupture energy. The data presented in this document can be used to improve the fire resistance of the existing materials studied.

1. Introduction

The fiberglass has been constituted in an engineering material widely used to improve the mechanical resistance of polymeric materials [1, 2]. The glass fiber reinforced polymer matrix composite has been used in the manufacture of various products due to its excellent physical and mechanical properties. It has been used in the automotive, aeronautical and naval industries as a coating for sports products, in the construction industry in devices such as skylights and decorative elements, as well as in various sports equipment. However, the flammable capacity of polymeric resins induces a potential fire risk for those products. Therefore, in recent years the concern of researchers in materials is the improvement of flame retardancy [3-5].

Industries such as the automotive, aeronautical and nautical industries make great efforts to improve the flame behavior of the composite materials used in the manufacture of their products. Positive results have been found with the use of flame retardants. Flame retardants are chemical substances that are added to polymeric compounds generally in the form of particles (powders) in controlled weight ratios. Several researchers have tested a series of organic and inorganic flame retardants to reduce the flammability of polymeric compounds [6]; In addition, they seek to determine the effect of these on the final mechanical properties [7, 8].

A composite material is a combination of two or more primary materials (metallic, non-metallic, ceramic and plastic). Generally, one of these materials is constituted as a matrix and another as a reinforcement. In glass fiber reinforced polymer composites, the matrix is made up of resin (Epoxy, polyester, etc.) and reinforcing or reinforcing material woven or small cross-linked glass fibers, manufactured by different techniques. Different percentages by weight of the reinforcement material and its shape affect the properties the physical-mechanical properties of the compound [9-12].
Studies have been reported on the positive influence of flame retardants based on aluminum hydroxide and magnesium hydroxide [13-16]; results of these studies showed that the flame resistance of the compounds improved significantly with the addition of said retardants, without considerable adverse effect on the final mechanical properties [8, 13, 17, 18]. One of the advantages over the use of these retardants is that during exposure to the flame, water and non-flammable gases are released, so that the increase in the propagation of the fire is reduced. In addition, they can also absorb heat through reactions that release gas, which produces some cooling. These advantages, among others, have been the reasons for its very common use in many applications [19, 20].
The purpose of this study is to design an experiment to investigate the use of aluminum hydroxide and magnesium hydroxide as filler material to manufacture fiberglass reinforced epoxy and polyester polymeric composite materials; at the same time detect their effects on the mechanical properties.

2. Experimental development

2.1. Design of experiments

An experimental design was established by a factorial design 3^3 (Table 1). Two qualitative variables (type of resin, type of retardant) and a quantitative variable (volumetric composition of the retardant) were used, at two and three levels respectively. Through all possible combinations obtained from the DOE, 5 mechanical properties of the composite material were evaluated: tensile strength (TS), tensile modulus (TM), flexural strength (FS), flexural modulus (FM) and impact resistance (IR). According to specific regulations, 5 replicas were executed for each experimental combination; obtaining 60 experimental runs for each property.

Table 1. Factors and levels of experimental design.

Factors	**Low**	**medium**	**High**
Type of resin	Epoxy	Polyester	-
Type of retardants	Aluminum hydroxide	Magnesium hydroxide	-
volumetric composition of the retardant	3%	6%	9%

2.2. Preparation of the composite material

For the elaboration of the compound were used: polyester resin series Reichold 33200 Polylite and Repox 6090A and epoxy resin brand QcuadCarbon. The reinforcement material was fiberglass type EMC 450 brand Taishan Fiberglass from China. Retardants: Extra pure aluminum hydroxide powder (Fisher Scientific brand) and magnesium hydroxide (grade 95% reagent, Sigma-Aldrich brand). The compound was manufactured by manual stratification, the overall volumetric composition is 70% matrix and 30% reinforcement. The formulations of the compounds are presented in Table 2.

Table 2. Volumetric composition of the composite material.

	Composite (%)											
Material combinations	**1**	**2**	**3**	**4**	**5**	**6**	**7**	**8**	**9**	**10**	**11**	**12**
Fiberglass	30	30	30	30	30	30	30	30	30	30	30	30
Epoxy Resin	67	64	61	67	64	61	-	-	-	-	-	-
Polyester resin	-	-	-	-	-	-	67	64	61	67	64	61
Aluminum hydroxide	3	6	9	-	-	-	3	6	9	-	-	-

Magnesium hydroxide	-	-	-	3	6	9	-	-	-	3	6	9
Total	100	100	100	100	100	100	100	100	100	100	100	100

The composition of the retardant is part of the 70% matrix. No reports have been found on minimum ranges established for the addition of flame retardants in composites of resinous matrix, so it was based on compositions smaller than those used by [21].

2.3. Mechanical characterization

The mechanical tensile, flexural and impact tests were followed by the instructions specified in the ASTM D3039-08, ASTM D7264M-07 and ASTM D5628-10 standards respectively; about the preparation of test tubes and testing.

3. Results and discussion

3.1. Experimental analysis

The experimental design was analyzed in Minitab18.1 Statistical Software. Table 3 shows the average of the five replications of the mechanical properties RT, MET, RF, MEF and ER evaluated by destructive tests under previously determined regulations.

Table 3. Average values of Ra and Fd for AGFRP y CGFRP.

N°	RESINE	RET	% RET	RT	MET	RF	MEF	ER
1	Epoxy	HA	3	72,65	2037,13	196,78	5794,06	2,53
2	Epoxy	HA	6	85,32	2030,61	152,45	4495,37	2,73
3	Epoxy	HA	9	79,35	2161,81	131,85	3754,27	4,00
4	Epoxy	HM	3	60,37	1819,70	134,41	3831,63	4,42
5	Epoxy	HM	6	56,56	1465,06	161,54	5082,48	2,98
6	Epoxy	HM	9	57,34	1710,30	154,34	4481,84	3,56
7	Polyester	HA	3	52,96	1980,84	163,63	4892,47	3,35
8	Polyester	HA	6	80,09	2305,82	206,41	7518,99	3,20
9	Polyester	HA	9	92,08	2801,39	141,12	5044,56	1,70
10	Polyester	HM	3	41,82	1715,90	150,18	4763,11	2,57
11	Polyester	HM	6	52,67	2309,97	254,09	10705,51	2,61
12	Polyester	HM	9	90,68	2670,32	198,07	6860,84	2,25

In Figure 1 we show the main effects of the factors on the output variable (mechanical properties). Regarding RT (Figure 1a); none of the resins offers a maximization, so any of the 2 could be used. the retardant that allows obtaining the highest value of RT is HA. Also, it is indicated that for there to be a maximization of the value of RT it is necessary to use 9% of flame retardant. In the case of FT (figure 1b), the resin that offers a maximization for FT is polyester. On the other hand, the flame retardant that allows to obtain a better behavior to the flexion is the HM, adding itself in volumetric composition of 6%. With respect to ER (Figure 1c), the resin that allows obtaining the highest value of ER is epoxy. The inclusion of the flame-retardant HM allows obtaining the maximum ER by adding a 3% volumetric fraction.

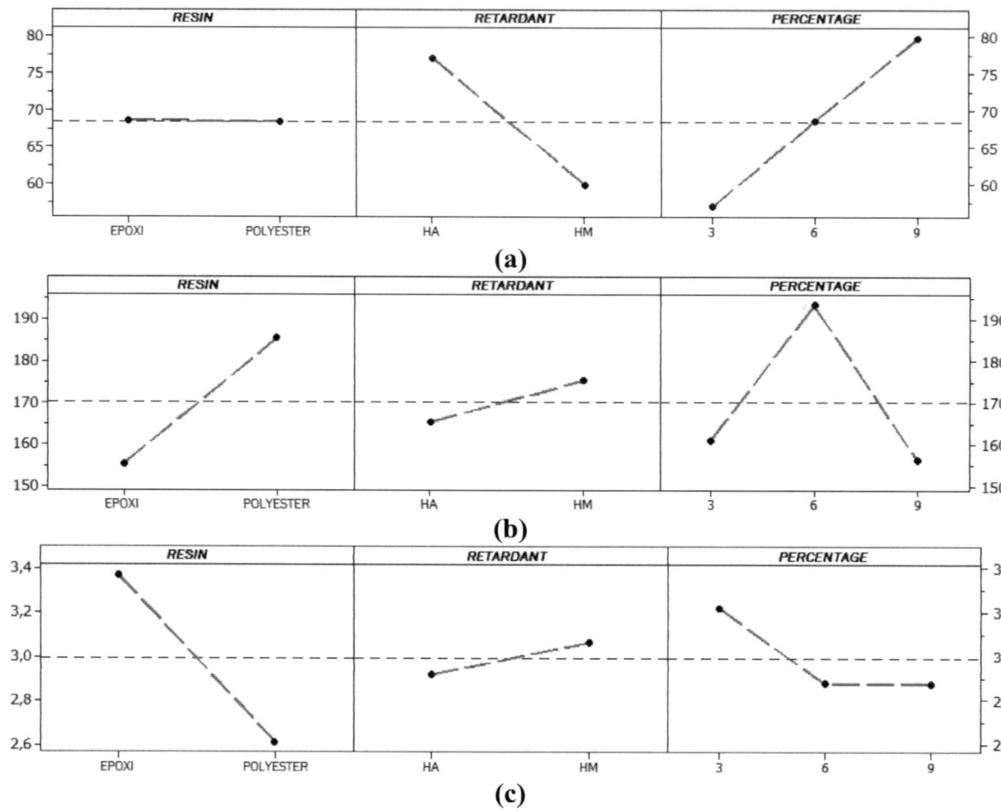

Figure 1. Main effects of factors on mechanical properties. **a)** Tensile strength, **b)** Flexural strength, **c)** Rupture energy.

Figure 2 shows the interactions of the factors and how they affect the mechanical properties RT, FT and ER. In general, the compound with polyester resin shows a better performance of RT and FT using any flame retardant in percentages of 6 and 9%. In contrast, the epoxy matrix compound shows less affectation to its tenacity (ER).

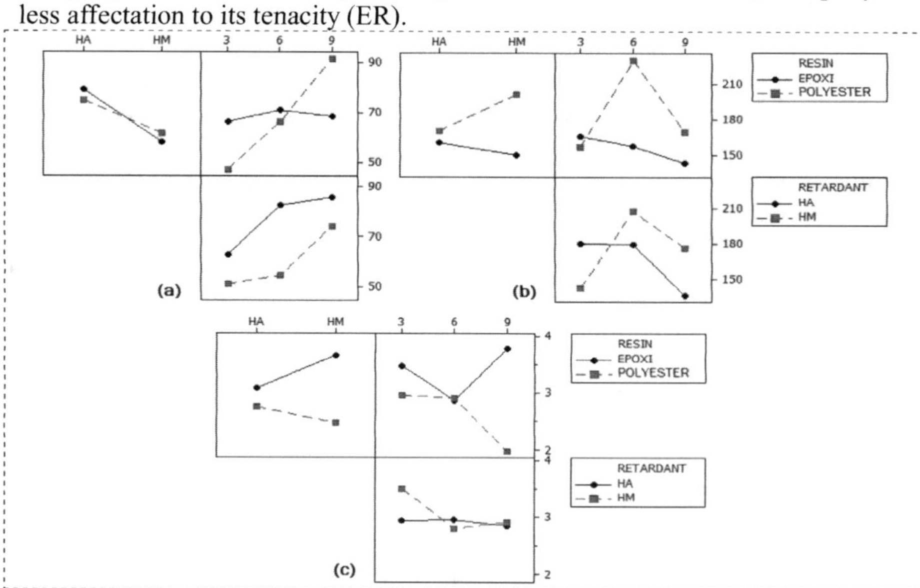

Figure 2. Interactive effects of factors on mechanical properties. **a)** Tensile strength, **b)** Flexural strength, **c)** Rupture energy.

The flame-retardant HA in volumetric composition of 9% shows the least affectation to RT of the compound with polyester matrix (Figure 2a). In the case of FT, the best configuration was: polyester matrix stratified with 6% of HM (Figure 2b). Finally, the tenacity of the compound had less affectation when 3% of the HM retarder was occupied in the epoxy matrix; however, when increasing the percentage of retardant, HA shows less influence than HM (Figure 2c).
The global effects are shown in figure 3. For both RT and RF (Figure 3a and 3b respectively), the application of retardant with different addition percentages influences resistance, not so much in ER (figure 3c), where its influence is not significant. This is supported by an ANOVA analysis performed in parallel to the experimental analysis with an adjustment of 61% and with 95% confidence. Although the value of the obtained adjustment was not very high, this gives us an expectation to the experimental continuation with greater replicas and a study in greater depth with each of the fire retardants.

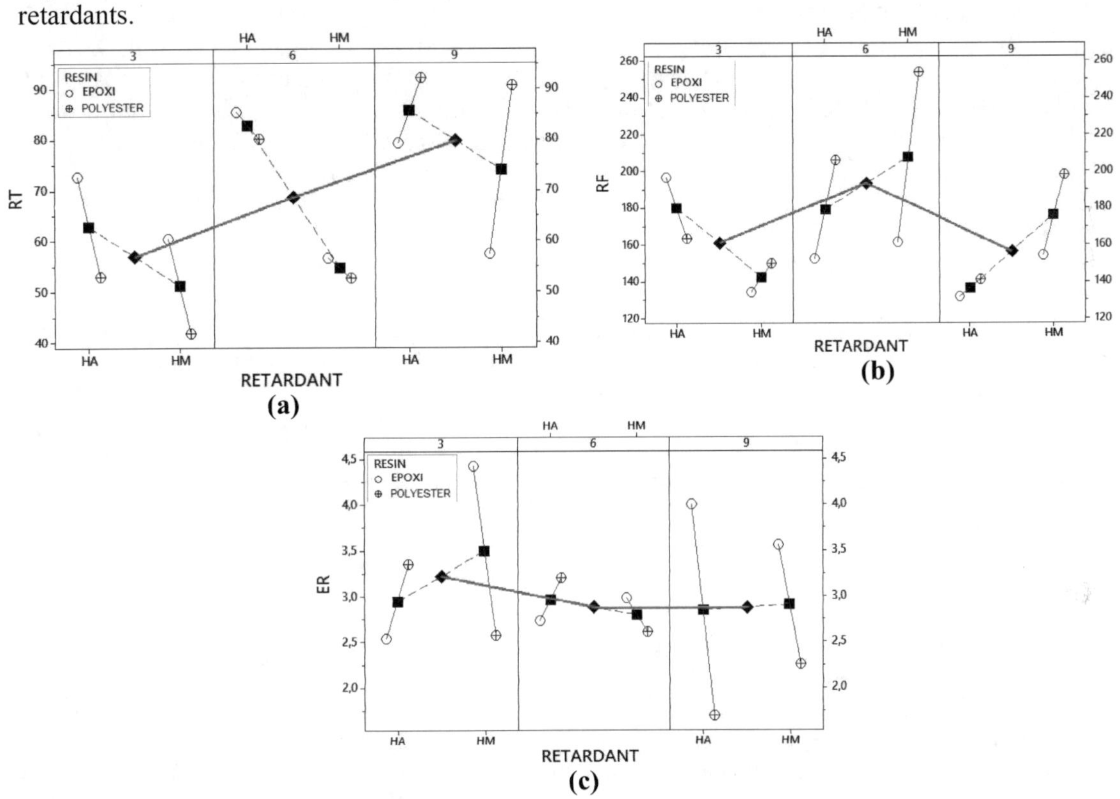

Figure 3. Effects of experimental factors on mechanical properties. **a)** Tensile strength, **b)** Flexural strength, **c)** Rupture energy.

For the contrast of the results obtained, mechanical tests were carried out on the materials without the addition of retardants, which were shown in Table 4. Compared with the results obtained (Table 3), the properties were affected by the addition of the retarders in the matrix is minimal; and in some cases, improves (combination 11 and 12 for RT and MET, 1,8,11 and 12 for RF and MEF).

Table 4. Mechanical properties of the compound without retardant combination

Description of material and different Matrix	Tensile strength, ultimate (MPa)	Tensile modulus (MPa)	Flexural strength (MPa)	Flexural modulus (MPa)	rupture energy (J)
Composite - Epoxy	83,6	2258	157,4	4331	4,55
Composite - Polyester	85,5	2750	155,2	5410	3,72

Comparing the results with [13], which indicates that the tensile and flexural strength of glass fiber composite materials with polyester resin with 3% and 6% aluminum hydroxide is lower compared to the pure composite material without retarder, so it establishes that the addition of aluminum hydroxide had a negative impact on tensile and flexural properties; however, in the present study it was determined that the addition of flame retardants does not necessarily have a negative effect on the mechanical properties. For the impact test on the rupture energy a small decrease of this property was evidenced for all the configurations of the material. These results, in turn, contrast those determined by [22], in which it indicates that the effect of the content of flame retardants in the composite materials with Polypropylene was significant. The result of said study obtained an increase in impact resistance for the compounds of 13% compared to those without retarder; therefore, the cause and effect of each type of retardant and its percentage of addition on the different compounds in their original material should be investigated in the future.
Finally, it can be deduced that the application of these HA and HM organic retardants have an acceptable efficiency (the mechanical properties do not vary significantly) and especially because the hydroxides have the reputation of being non-toxic because they are friendly to the environment. In contrast to this, halogen-based flame retardants release toxic smoke, which increases the concentration of poisonous and corrosive gases in the fire, which can easily lead to suffocation.

4. Conclusions
This study shows that the use of flame retardants in polymer matrix composite materials affects the mechanical properties.
The improvement of: RT, MET (obtained in the combination of group 9) and FT, MEF (obtained with group 11) should be investigated in depth in future studies, since they showed higher values than those obtained without adding retardant.
Finally, it is evident that the combinatorial effects of the experimental design factors have different effects on the mechanical properties, so it is recommended to use the best configurations depending on the resistance requirements in specific applications.

5. References
[1] J. Singh, M. Kumar, S. Kumar, and S. Mohapatra, "Properties of glass-fiber hybrid composites: a review," *Polymer-Plastics Technology and Engineering,* vol. 56, pp. 455-469, 2017.
[2] T. Sathishkumar, S. Satheeshkumar, and J. Naveen, "Glass fiber-reinforced polymer composites–a review," *Journal of Reinforced Plastics and Composites,* vol. 33, pp. 1258-1275, 2014.
[3] J. Jiang, Y. Cheng, Y. Liu, Q. Wang, Y. He, and B. Wang, "Intergrowth charring for flame-retardant glass fabric-reinforced epoxy resin composites," *Journal of Materials Chemistry A,* vol. 3, pp. 4284-4290, 2015.
[4] B. Z. Dholakiya, "Use of non-traditional fillers to reduce flammability of polyester resin composites," *polimeri,* vol. 30, pp. 10-17, 2009.
[5] S. Nazare, B. Kandola, and A. Horrocks, "Flame - retardant unsaturated polyester resin incorporating nanoclays," *Polymers for advanced technologies,* vol. 17, pp. 294-303, 2006.
[6] A. Kausar, I. Rafique, Z. Anwar, and B. Muhammad, "Recent developments in different types of flame retardants and effect on fire retardancy of epoxy composite," *Polymer-Plastics Technology and Engineering,* vol. 55, pp. 1512-1535, 2016.
[7] X. Zhao, L. Yang, F. H. Martin, X.-Q. Zhang, R. Wang, and D.-Y. Wang, "Influence of phenylphosphonate based flame retardant on epoxy/glass fiber reinforced composites (GRE): flammability, mechanical and thermal stability properties," *Composites Part B: Engineering,* vol. 110, pp. 511-519, 2017.
[8] O. C. Gunes, R. Gomek, A. Tamar, O. K. Kandemir, A. Karaorman, and A. Z. Albayrak, "Comparative Study on Flame Retardancy, Thermal, and Mechanical Properties of Glass Fiber

Reinforced Polyester Composites with Ammonium Polyphosphate, Expandable Graphite, and Aluminum Tri-hydroxide," *Arabian Journal for Science and Engineering,* pp. 1-8, 2018.
[9] D. D. Chung, *Composite materials: science and applications*: Springer Science & Business Media, 2010.
[10] F. Matthews and R. Rawlings, "10-Short fibre composites," *Composite Materials,* pp. 287-325, 1999.
[11] C. Perez, J. Paredes, E. Lalaleo, C. Arroba, and D. Nuñez, "ANÁLISIS NUMÉRICO/EXPERIMENTAL DE LAS PROPIEDADES MECÁNICAS DEL COMPUESTO DE MATRIZ POLIÉSTER REFORZADO CON CASCARILLA DE ARROZ," *Ciencia,* vol. 19, 2018.
[12] C. Pérez, F. Palacios, H. Vaca, J. Paredes, and D. Núñez, "Mechanical characterization of a composite material of latex elastomeric matrix reinforced with recycled rubber granules to obtain floors," in *AIP Conference Proceedings*, 2018, p. 020011.
[13] P. Hiremath, H. Arunkumar, and M. Shettar, "Investigation on Effect of Aluminium Hydroxide on Mechanical and Fire Retardant Properties of GFRP-Hybrid Composites," *Materials Today: Proceedings,* vol. 4, pp. 10952-10956, 2017.
[14] A. Frańczak, M. Oleksy, R. Oliwa, and G. Budzik, "Polyethylene composites flame retarded with aluminum hydroxide as coatings for electrical cables," *Polimery,* vol. 63, 2018.
[15] H. Gui, X. Zhang, Y. Liu, W. Dong, Q. Wang, J. Gao, *et al.*, "Effect of dispersion of nano-magnesium hydroxide on the flammability of flame retardant ternary composites," *Composites science and technology,* vol. 67, pp. 974-980, 2007.
[16] M. Wang, X.-F. Zeng, J.-Y. Chen, J.-X. Wang, L.-L. Zhang, and J.-F. Chen, "Magnesium hydroxide nanodispersion for polypropylene nanocomposites with high transparency and excellent fire-retardant properties," *Polymer Degradation and Stability,* vol. 146, pp. 327-333, 2017.
[17] M. R. Petersen, A. Chen, M. Roll, S. Jung, and M. Yossef, "Mechanical properties of fire-retardant glass fiber-reinforced polymer materials with alumina tri-hydrate filler," *Composites Part B: Engineering,* vol. 78, pp. 109-121, 2015.
[18] P. Galande and S. Zarekar, "Effect of Various Fillers on Mechanical Properties of Glass Fiber Reinforced Polymer Composites: A Review."
[19] F. Scientific. (2019). *Aluminum hydroxide - Specifications Sheets*. Available: https://www.fishersci.com/shop/products/aluminum-hydroxide-powder-acros-organics-4/AC219130250
[20] SIGMA-ALDRICH. (2019). *Magnesium hydroxide - Specifications Sheets*. Available: https://www.sigmaaldrich.com/catalog/product/sigald/310093?lang=es&region=MX
[21] J.-Z. Liang and F.-J. Li, "Impact fracture toughness and morphology of polypropylene/Mg (OH) 2 composites," *Journal of Polymer Engineering,* vol. 33, pp. 721-725, 2013.
[22] J. Liang, J. Feng, C. Tsui, C. Tang, D. Liu, S. Zhang, *et al.*, "Mechanical properties and flame-retardant of PP/MRP/Mg (OH) 2/Al (OH) 3 composites," *Composites Part B: Engineering,* vol. 71, pp. 74-81, 2015.

Acknowledgment
The authors thank the DIDE of the "Universidad Técnica de Ambato" for the total support in the execution of this study "Análisis y determinación de índices de inflamabilidad de materiales utilizados en la fabricación del interior de carrocerías metalmecánicas de la ciudad de Ambato" PFICM007.

IOP Conf. Series: Materials Science and Engineering **628** (2019) 012002 doi:10.1088/1757-899X/628/1/012002

Influence of steps temperature on microstructure and hardness of TiO_2 thin films deposited by co-sputtering

L García González[1], J B Santaella González[1], F López Huerta[2,*], E. Díaz Trueba[2], L Zamora Peredo[1], C Zuñiga Islas[3] and C Guarneros Aguilar[4]

[1] Centro de Investigación en Micro y Nanotecnología, Universidad Veracruzana, Veracruz, 94294, Boca del Río, Veracruz, México.

[2] Facultad de Ingeniería Eléctrica y Electrónica, Universidad Veracruzana, Veracruz, Boca del Río, Veracruz, México.

[3] Departamento de Electrónica, Instituto Nacional de Astrofísica, Óptica y Electrónica (INAOE), C.P. 72840 Tonantzintla, Puebla, México.

[4] CONACYT, Instituto Politécnico Nacional, CICATA, 89600, Altamira, Tamaulipas, México.

*e-mail: frlopez@uv.mx

Abstract. TiO_2 thin films were synthetized by magnetron co-sputtering on silicon (100) substrates. Two targets were used: Ti and TiO inside an inert atmosphere of Argon gas at room temperature. Post-deposition thermal annealing treatments were performed with different numbers of step temperature segments to reach 500 °C for 3 hours. The effect of the variation of the number of step temperature segments in the micro hardness and microstructure of the films was studied, using XRD, Raman spectroscopy, SEM, AFM, photoluminescence (PL) and Vickers hardness. Vickers hardness evinced values between 14.7 and 19.8 GPa. Raman spectroscopy showed that the film without thermal annealing does not present any active bands, while films with post deposit annealing treatments had rutile and anatase phases with higher intensity as the number of step temperature segments increased, this behavior was corroborated by XRD. SEM and AFM showed a change in the morphology as the number of step temperature segments increased. Moreover, PL showed that oxygen defects decrease as the annealing time increases; which could be related to the change in hardness and morphology.

1. Introduction

Titanium dioxide (TiO_2) has three crystalline phases, anatase, rutile and brookite [1], which give it great versatility in applications such as gas sensors, photocatalysis and protective material [2]. Recent investigations mention various ways of synthesizing TiO_2 thin films, wherein they can be deposited by diverse processes, such as sol-gel, electrodeposition, pulse laser deposition (PLD) or Sputtering, among others. Magnetron Sputtering is used due to its advantages of reproducibility, film composition which is relatively easy to control, good adhesion to the substrates, a controllable thickness, and the possibility to scale the fabrication of the films at industrial levels [3- 4]. The formation of TiO_2 films has varying certain parameters during the sputtering process such as, bias voltage [1,3], source power supplies [5], or substrate temperature to favor the presence of a crystalline phase [6] and with the atmosphere of argon + oxygen [7]; which has been reported in several studies.

However, only a few studies allude to post-deposit annealing treatments that form a crystalline phase, such as the anatase and rutile phases [8-9], or mention the application of step temperature segments (dwell segments), and study the effect it has on its hardness. Therefore, we carried out studies to analyze the way in which the heating conditions, like step temperature segments, in post-deposit annealing treatments can improve the formation of the TiO_2 crystalline phases, as well as the consequences these changes have on their hardness, microstructure and crystalline defects.

2. Methodology

2.1. Substrate preparation and coating synthesis

TiO_2 thin films were deposited onto 4-inch Silicon (100) wafers using a sputtering system, Intercovamex V3, where two targets were used: Ti (99.995%) in the DC source with a power of 120 W and TiO (99.99%) in the RF source with a power of 40 W. Formerly, the silicon wafers were cleaned using the RCA method [10]. In the co- Sputtering deposition process, co-sputtering consists of starting the two targets simultaneously to carry out the deposition process of TiO_2 thin films for 20 minutes and was performed at ambient temperature using an inert atmosphere of ultra-high purity Argon gas (99.999%) at a flow rate of 15 standard cubic centimeters per minute (sccm), the pre-deposit vacuum was 2.0×10^{-6} Torr.

After the films were fabricated, post-deposition thermal annealing treatments (TT) were performed at a maximum temperature of 500 °C for 3 hours in air, varying the number of step temperature segments to decrease the heating rate to reach the maximum temperature. Three different arrangements were used: one-step temperature segment at 250°C (labeled as 1RTT) with a duration of 30 minutes, two-step temperature segments at 167 and 334 °C for 30 minutes for each of the steps (2RTT), and three-step temperature segments at 125, 250 and 375°C for 30 minutes for each of the steps (3RTT). Finally, after completing the temperature steps, the temperature of 500 ° C was reached and stabilized, which was maintained for 3 hours, and then it was allowed to cool inside the muffle until it reached room temperature.

2.2. Characterization techniques

The vibrational excitations of the TiO_2 thin films were obtained by Raman spectroscopy using a DXR Raman Microscope with a 532-nm laser and with a power of 10 mW. The Raman measurements were performed at room temperature. The crystal structures were analyzed with a grazing angle configuration X-ray diffractometer, Bruker D8 Advance, using CuK_α radiation at a scanning speed of 1°/min. The diffractograms were analyzed using the Scherrer method to compute the grain size [11] and additionally, the semi-quantitative analysis of phases was analyzed in the EVA software from BRUKER. To observe the photoluminescence characterization a system is formed by, a 325-nm He-Cd Kimmon laser, a chopper with a frequency of 37 Hz, and a Merlin Radiometry Systems model 70103 were used. The thickness and morphology of the films was measured in a JEOL JSM-7600F field emission scanning electron microscope (FE-SEM) using an acceleration voltage of 3 kV and a working distance of 3.0 mm. Moreover, Atomic Force Microscope (AFM) was used to analyze the topography of the films, using a Nanosurf EasyScan in dynamic mode, with a non-contact tip type and a constant force of 48 N/m. Finally, the hardness measurements were performed with a Mitutoyo Micro Vickers HM-125 hardness testing machine; using load values from 2, 3, 4, 5, 10, 25, 50, 100, 200 and 300 gram-force (gf), was applied load they were repeated five times and then the Korsunsky work of indentation approach [12], whereby despite the contribution of the hardness of the silicon substrate, the adjustment consists of being able to extract the hardness value of the thin film.

3. Results and discussion

3.1 Microstructure

After the microstructure was studied by X-ray diffraction of the thin films of TiO_2 with thermal annealing, where the diffractograms show well defined peaks in addition to the silicon peaks which correspond to the anatase and rutile (TiO_2) phases. There, it can be seen than the peaks of the rutile are more intense than the anatase peaks, which could mean that the films are a combination of rutile and anatase, but also, it means that there is a higher percentage of rutile peaks, which corresponds to an EVA analysis that shown in Table 1. This effect is not quite noticeable in Raman, since the bands of the rutile phase are near the anatase phase, and the rutile bands are usually less intense[13-14].
Then Raman spectroscopy was studied, an evolution to a better crystalline quality is seen in Figure 1.b), where the bands at 300 and 530 cm $^{-1}$ correspond to the Silicon substrate (100), There are also well-defined bands that correspond to the anatase phase [15], where the band formation at ~144 cm^{-1} corresponds to the E_g vibrational mode, and the bands at about ~398 cm^{-1} and ~639 cm^{-1} correspond to the B_{1g} and E_g vibrational modes of anatase respectively [13]. Moreover, an increment in the intensity of these anatase bands is seen from the 1RTT to the 2RTT, but the hop in intensity is not visible between 2RTT or 3RTT; apparently a better anatase formation occurs as the number of steps increase. But, according to other Raman analyses of the same material with a mixture of crystalline phases, the peak over the 144 cm^{-1} corresponding to the crystalline phase anatase is always much more intense even if there is presence of rutile in greater proportion [14]. Additional poorly formed bands are seen in Figure 1.b), at about ~238 cm^{-1} (Eg), ~448 cm^{-1} (Eg) and another at ~615 cm^{-1} (A1g) [14], which might be related to the rutile phase, since it is well known that these phases can coexist [13], this is corroborated in Figure 1 a), where the diffractograms show well defined peaks.

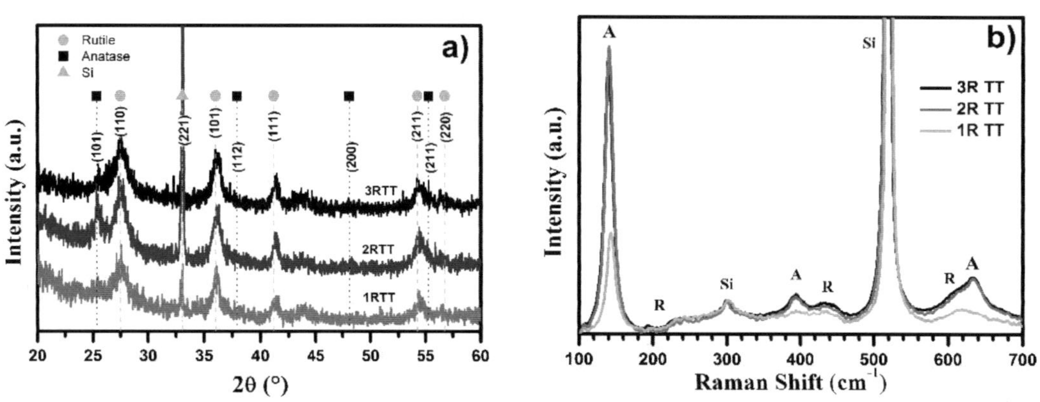

Figure 1. Microstructure of the thin films studied by a) X-ray diffractogram of the TiO_2 thin films with thermal annealing, and b) the Raman spectra of the thin films with thermal annealing.

In Table 1 the XRD results of the films are summarized, the rutile phase mainly appears when the thermal annealing is made with one-step temperature segment (1RTT), this could be related to the synthesis process, since other works report that the Sputtering technique tends to form rutile or disordered TiO_2, then a thermal annealing rutile form [16]. Nonetheless, increasing the number of steps reduces the rutile phase percentage and increases the anatase phase percentage, where the percentage values are like the 2RTT or 3RTT, in agreement with the behavior seen in Raman spectra intensities. However, the three-step temperature segment film tends to decrease the grain size of the rutile, while the anatase remains relatively the same. Due to the use of steps in temperatures lower than 500 ° C, where the formation of anatase is favored, this is why this phase begins to increase with the addition of steps [16]. This phase change could be explained by the low temperature values of the thermal annealing, due to the reported optimal temperature for anatase formation being between 400 and 500 °C [15], the fact that the rutile formed at that temperature depends more on the synthesis process rather than the thermal treatment [1,17].

Table 1. XRD phase percentage and crystal (tetragonal) size for the thin films with thermal annealing

	Rutile		Anatase	
	Phase %	Crystal Size (nm)	Phase %	Crystal Size (nm)
1R-TT	82.66%	20.4	17.34%	28.2
2R-TT	68.4%	11.8	31.6%	37.1
3R-TT	69.67%	8.7	30.33%	31.5

3.2 Photoluminescence and AFM

The photoluminescence spectrum is shown in Figure 2 a) in the energy range of 2.0 and 3.5 eV and shows three main peaks: one at 3.2 eV related to the Band Gap of the TiO_2 and another two peaks at 2.1 and 2.8 eV, which could be related to oxygen defects [18]. The Band Gap peak does not change between the films with different step temperature segments, neither in position nor intensity, which could indicate that while there is a microstructural change, the band gap remains relatively the same. On the other hand, the peaks related to the oxygen defects tend to diminish as the number of steps increases; this behavior could be attributed to a better crystalline quality shown in the microstructural studies, as it can be observed that they have the same behavior with Raman Figure 1 b) and in X-ray diffractogram Figure 1 a), since the slow increment in temperature allowed the oxygen atoms to have more time to rearrange into a more stable structure [8]. Using Gwyddion software, the surface images of the thin films were obtained, where the RMS roughness values could be calculated in an area of 1 μm^2. In Figure 2 b), the AFM images show the surface topography of the thin film with 1RTT, where all films evinced a good homogenization with a surface than deepened less than 10 nm. The rugosity values are depicted of the films, there, the film without heat treatment (WTT) has an RMS rugosity of 1.25 nm, 1RTT films 1.30 nm, 2RTT 1.42 nm, and this increases as the number of step temperature segments reaches 1,68 nm for 3RTT. This may be due to the increased crystallinity level of the films, as observed in the Raman and XRD data, as they have shown in some articles [1].

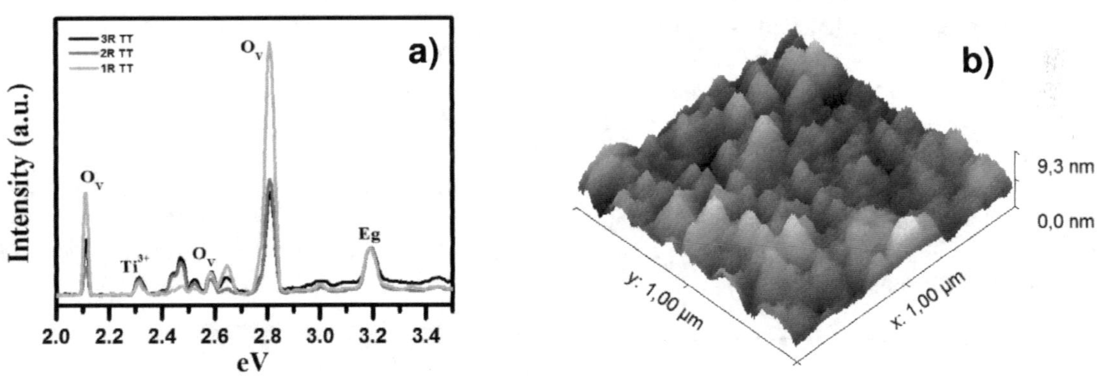

Figure 2. a) Photoluminescence spectrum of the thin films with thermal annealing in the range from 2.0 to 3.5 eV, and b) Topography of the thin films: 1RTT.

3.3 Morphology

Figure 3 shows surface morphology without thermal annealing and 1R TT, in which a relatively large area is observed and where a very uniform surface is observed in images a) y b), the other samples show a similar morphology, so they were not placed. The formation of two different sizes of nanometric grain can be seen, one around 30 nm and the other approximately at 15 nm, all very similar and uniform sizes on the surface of the respective thin films. This corroborates the grain sizes calculated in the XRD analysis of Table 1, where two different crystalline phases are identified. In all the SEM images of the surface of the thin films, the morphology shown of the material does not shows apparent changes regarding what has been reported in several studies [16].

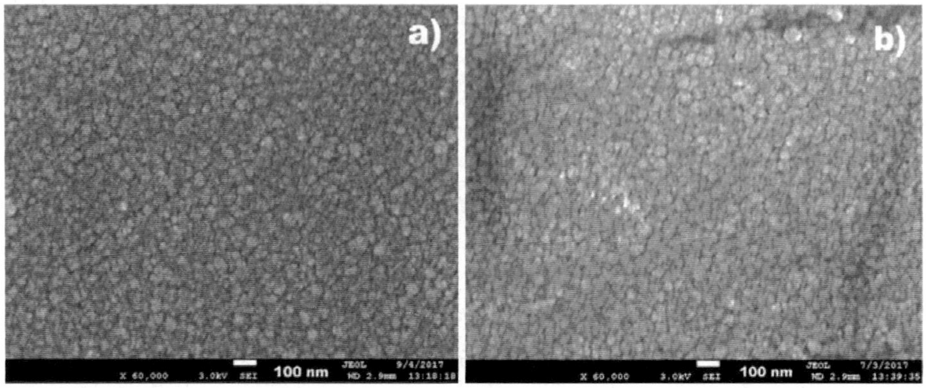

Figure 3. Image of the SEM where morphology of the surface samples can be observed: a) films without thermal treatment (WTT), and b) 1RTT.

3.4 Hardness and thickness.

Figure 4 a) shows the comparison of hardness vs rugosity of the films with different TT; the hardness values were adjusted applying the Korsunsky work-of-indentation approach model, which is a method of extracting a value of hardness [12]. Results of the Vickers Microhardness showed a clear behavior that increased the hardness values as the number of steps increased, reaching hardness values of 14.74, 18.09 and 19.82 GPa for the 1RTT, 2RTT and 3RTT respectively. This enhanced hardness of the films at higher temperature steps could be related to the reduction of the grain size of the rutile phase seen in the XRD results [13], additionally, it is possible that the hardness increase could be because they have better crystalline quality, as can be verified with Raman spectra and Photoluminescence. While hardness above 18 GPa could seem too high for TiO_2 two films, other studies have archived similar results [19-20] and some similarities can be observed between the reported works and our work, the grain sizes are very similar, and is also influenced by the type of substrate (silicon) [21], that have reached hardness values of 19 GPa. It is visible that the effect of the step temperature segments is favorable for improving the hardness and crystal quality on the rutile due to the reduction in the crystal size that is shown in Table 1. This could be related to the fact that when steps were used, a greater homogeneity of the temperature is ensured within the furnace, and therefore the thin films, before reaching the final temperature of the thermal treatment. Finally, the thickness was measured in cross section, as shown in Figure 4 b), the micrograph of the cross section of the 1RTT film, where it shows a thickness of 400 nm, where a type of columnar growth of the material is observed, the thickness of thin films does not vary.

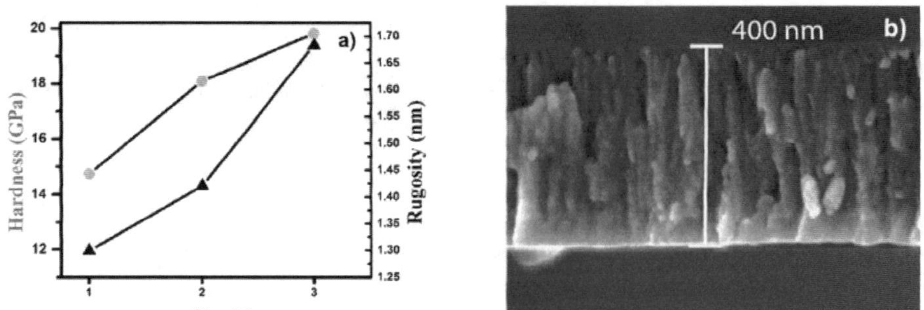

Figure 4. a) Vickers measurements plot of hardness and rugosity vs steps in thermal treatment, and b) cross- sectional SEM of the 1RTT thin film.

IOP Conf. Series: Materials Science and Engineering **628** (2019) 012002 doi:10.1088/1757-899X/628/1/012002

4. Conclusions

Thin films of TiO_2 were fabricated by co-sputtering technique on silicon wafer (100) substrates and post- deposition thermal annealing treatments were made at 500 °C for 3 hours, varying the number of steps to reach maximum temperature. The thin film with the best crystalline quality corresponds to the 3RTT film, this is demonstrated by the Raman and PL spectrum. For hardness and rugosity, this film reached maximum hardness values of 19.8 GPa, AFM revealed the highest RMS roughness value to be 1,68 nm in its topography, being the best values of all films. It has been established that the crystalline quality, hardness and rugosity improve as the number of temperature steps increases, but the band gap does not change with it. With the proposed conditions in this study to implement the TiO_2, thin films have reached hardness values similar to 19 GPa compared to those reported in other studies between 16 GPa and 19 GPa. In other hardness studies, Vickers have reported similar hardness values for this material, TiO_2 shows a similar hardness for microhardness measurements or Vickers hardness.

5. References

[1] S. Nezar, N. Saoula, S. Sali, M. Faiz, "Properties of TiO_2 thin films deposited by rf reactive magnetron sputtering on biased substrates," *Appl. Surf. Sci.*, vol. 395, pp. 172–179, 2017.

[2] N. Kawashima, Q. Y. Zhu, and A. R. Gerson, "The selective growth of rutile thin films using radio- frequency magnetron sputtering," *Thin Solid Films*, vol. 520, no. 11, pp. 3884–3891, 2012.

[3] L. Bait, L. Azzouz, N. Madaoui, and N. Saoula, "Influence of substrate bias voltage on the properties of TiO_2 deposited by radio-frequency magnetron sputtering on 304L for biomaterials applications," *Appl. Surf. Sci.*, vol. 395, pp. 72–77, 2017.

[4] G. Varshney *et al.*, "Nanoscale TiO_2 films and their application in remediation of organic pollutants," *Coord. Chem. Rev.*, vol. 306, no. P1, pp. 43–64, 2016.

[5] L. Skowronski, B. Wicher, K. Zdunek, R. Chodun, S. Okrasa, and K. Nowakowska-Langier, "Optical TiO_2 layers deposited on polymer substrates by the Gas Injection Magnetron Sputtering technique," *Appl.* Surf. *Sci.*, vol. 466, no. September 2018, pp. 12–18, 2019.

[6] D. S. R. Krishna and Y. Sun, "Thermally oxidised rutile-TiO_2 coating on stainless steel for tribological properties and corrosion resistance enhancement," *Appl. Surf. Sci.*, vol. 252, no. 4, pp. 1107–1116, 2005.

[7] J. A. P. Nuñez, H. S. Salapare, M. M. S. Villamayor, M. A. T. Siringan, and H. J. Ramos, "Antibacterial efficiency of magnetron sputtered TiO_2 on poly(methyl methacrylate)," *Surfaces and Interfaces*, vol. 8, no. August 2016, pp. 28–35, 2017.

[8] B. Santara, P. K. Giri, K. Imakita, and M. Fujii, "Evidence for Ti interstitial induced extended visible absorption and near infrared photoluminescence from undoped TiO_2 nanoribbons: An in-situ photoluminescence study," *J. Phys. Chem. C*, vol. 117, no. 44, pp. 23402–23411, 2013.

[9] R. R. Guimaraes, A. L. A. Parussulo, and K. Araki, "Impact of nanoparticles preparation method on the synergic effect in anatase/rutile mixtures," Electrochimica Acta, vol. 22, pp. 1378-1386, 2016.

[10] R. Joyce, K. Singh, S. Varghese, and J. Akhtar, "Effective cleaning process and its influence on surface roughness in anodic bonding for semiconductor device packaging," *Mater. Sci. Semicond. Process.*, vol. 31, pp. 84–93, 2015.

[11] A. W. Burton, K. Ong, T. Rea, and I. Y. Chan, "On the estimation of average crystallite size of zeolites from the Scherrer equation: A critical evaluation of its application to zeolites with one-dimensional pore systems," *Microporous Mesoporous Mater.*, vol. 117, no. 1–2, pp. 75–90, 2009.

[12] A. M. Korsunsky, M. R. McGurk, S. J. Bull, and T. F. Page, "On the hardness of coated systems," *Surf.* Coatings *Technol.*, vol. 99, no. 1–2, pp. 171–183, 1998.

[13] W. Hui, S. Guodong, Z. Xiaoshu, Z. Wei, H. Lin, and Y. Ying, "In-situ synthesis of TiO_2 rutile/anatase heterostructure by DC magnetron sputtering at room temperature and thickness effect of outermost rutile layer on photocatalysis," *J. Environ. Sci.*, pp. 1–10, 2017.

[14] R. R. Guimaraes, A. L. A. Parussulo, and K. Araki, "Impact of nanoparticles preparation method on the synergic effect in anatase/rutile mixtures," *Electrochim. Acta*, vol. 222, pp. 1378–1386, 2016.

[15] M. Vishwas, K. Narasimha Rao, and R. P. S. Chakradhar, "Influence of annealing temperature on Raman and photoluminescence spectra of electron beam evaporated TiO_2 thin films," *Spectrochim. Acta- Part A Mol. Biomol. Spectrosc.*, vol. 99, pp. 33–36, 2012.

[16] G. Wan, S. Wang, L. Li, G. Mu, X. Yin, X. Zhang, Y. Tang, L. Yi, "Photocarrier dynamic measurement of rutile TiO_2 films prepared by RF magnetron reactive sputtering," *J. Alloys Compd.*, vol. 701, pp. 549–553, 2017.

[17] J. Lin, B. Wang, W. D. Sproul, Y. Ou, and I. Dahan, "Anatase and rutile TiO_2 films deposited by arc- free deep oscillation magnetron sputtering," *J. Phys. D. Appl. Phys.*, vol. 46, no. 8, p. 084008, 2013.

[18] A. Nemashkalo, "Photoluminescence and surface photovoltage studies of defects at nanoscale surfaces and interfaces in thin films of ZnO, TiO_2 and diamond," 2014.

[19] F. U. Lapostolle, A. Billard, and J. von Stebut, "Structure/mechanical properties relationship of titanium-oxygen coatings reactively sputter-deposited," *Surf. Coatings Technol.*, vol. 13, no.1, pp. 1–7, 2000.

[20] M. Birkholz, U. Albers, and T. Jung, "Nanocomposite layers of ceramic oxides and metals prepared by reactive gas-flow sputtering," *Surf. Coatings Technol.*, vol. 179, no. 2–3, pp. 279–285, 2004.

[21] V. Kulikovsky, R. Ctvrtlik, V. Vorlicek, J. Filip, P. Bohac, and L. Jastrabik, "Mechanical properties and structure of TiO_2 films deposited on quartz and silicon substrates," *Thin Solid Films*, vol. 542, pp. 91– 99, 2013.

Acknowledgements
J. B. Santaella González would like to thank the Micro and Nanosystems Postgraduate program and CONACYT for the scholarship granted and PRODEP 511-6/18-9245-UV-PTC-883. The authors acknowledge Leticia Tecuapetla from INAOE for AFM measurements, and Ing. Rebeca Rodríguez Jiménez for FE-SEM measurements.

IOP Conf. Series: Materials Science and Engineering **628** (2019) 012003 doi:10.1088/1757-899X/628/1/012003

Biocompatibility and surface properties of hydrogenated amorphous silicon-germanium thin films prepared by LF-PECVD

F López Huerta[1,*], R M Woo García[2], L García González[3], A L Herrera May[2,5], W Calleja Arriaga[3], R Vega[4], and E Soto[4]

[1] Facultad de Ingeniería Eléctrica y Electrónica, Universidad Veracruzana, Veracruz, Boca del Río, Veracruz, México.

[2] Centro de Investigación en Micro y Nanotecnología, Universidad Veracruzana, Veracruz, 94294, Boca del Río, Veracruz, México.

[3] Departamento de Electrónica, Instituto Nacional de Astrofísica, Óptica y Electrónica (INAOE), C.P. 72840 Tonantzintla, Puebla, México.

[4] Instituto de Fisiología, Benemérita Universidad Autónoma de Puebla, San Claudio, 72570, Puebla, Mexico.

[5] Maestría en Ingeniería Aplicada, Facultad de Ingeniería de la Construcción y el Hábitat, Universidad Veracruzana, Veracruz, Boca del Río, Veracruz, México.

*e-mail: frlopez@uv.mx

Abstract. We studied the surface morphology and biocompatibility of hydrogenated amorphous silicon-germanium (a-$Si_{1-x}Ge_x$:H) thin films prepared by Low Frequency Plasma Enhanced Chemical Vapor Deposition (LF-PECVD). These films were deposited on a Corning 2947 glass substrate having a thickness of 3 μm, the electrical performance showed a decreased electrical resistance for low regime voltage. The root mean square (RMS) surface roughness of the films was measured by atomic force microscopy (AFM) in a non-contact mode. A biocompatibility tests was carried out using primary cultures of dorsal root ganglion (DRG) of Wistar rats. The DRG neurons were incubated for 18 hours on a-$Si_{1-x}Ge_x$:H thin films, and subsequent electrophysiological recording was performed. These neurons displayed typical ionic currents, including a fast-inward current at the beginning of voltage clamp pulse (Na^+ current) and ensuing outward currents (K^+ current). In current clamp experiments, depolarizing current pulse injection caused typical action potential discharge of the neurons. These results confirmed the feasibility of using a-$Si_{1-x}Ge_x$:H thin films as a biocompatible material.

1. Introduction

Recent advances in integrated circuit (IC) and micro-electro-mechanical systems (MEMS) technologies allowed the development of new devices using unconventional materials, which can be used in environment pollution monitoring and biomedical applications [1-3]. Some of these materials, such as silicon germanium (SiGe), polysilicon SiGe, and boron doped amorphous silicon germanium (a-SiGe:B) can be obtained at low temperature of deposition (T <300 ºC). This has allowed the integration of different MEMS devices with standard complementary metal oxide semiconductor

(CMOS) electronics circuits [4-7]. Based on this integration, several devices have been fabricated such as pressure sensors, thin film membranes for cochlear implants, resonators for biomolecular sensing, micromirrors, tactile and tilt sensors, and surface ionization gas sensors [4-11]. In addition, MEMS devices can be integrated in standard CMOS process to obtain important advantages such as a single chip, monolithic integration, reduction of the parasitic effects, and low cost assembly and packaging [12-15]. An alternative material to fabricate MEMS devices with ICs is hydrogenated amorphous silicon-germanium film (a-$Si_{1-x}Ge_x$:H), which can be deposited on a different substrate by Low Frequency Plasma Enhanced Chemical Vapor Deposition (LF-PECVD) [16-17]. This material has suitable electrical and mechanical properties to fabricate MEMS devices. The properties of a-$Si_{1-x}Ge_x$:H films can be modified according to the type and ratio of the introduced gases during its deposition [16-17]. Generally, a-$Si_{1-x}Ge_x$:H films have been developed for applications in optoelectronic devices due to their optical band gap, which depends on the germanium content in the solid phase [17]. However, defining the biocompatibility of a-$Si_{1-x}Ge_x$:H films could allow for their potential application as biomaterials. Currently, it is difficult to obtain biomaterials with both good bulk and surface properties. a-$Si_{1-x}Ge_x$:H films have physical properties that could be suitable for advanced biological applications. However, more research about the biocompatibility of a- Sil-xGex:H films is necessary to guarantee its suitability as a biomaterial. A good biocompatibility of these films must avoid toxic or injurious effects on biological systems. Thus, biocompatibility tests must evaluate the biomaterial cytotoxicity and the functionality of the biological system when exposed to this biomaterial [18-19]. The MMT 3-(4,5-dimethylhiazol-2-yl)-2,5- diphenyl-2H-tetrazolium bromide) in vitro cytotoxicity assay is a convenient method for assessing cell viability [19] and to evaluate the cell viability, various parameters are used such as adhesion, spreading and recording of neural activity. Biocompatibility studies on an inorganic material require complex experiments both in vitro and in vivo in order to test the local and systemic effects of the material on cultured cells, tissue sections, and the whole body. In vitro cell culture tests are often used to screen the biocompatibility of inorganic material [20]. These tests are sensitive, reliable, convenient, and reproducible screening methods. The study of the biocompatibility in vitro offers a controlled environment to test specific cellular and molecular hypotheses. The tissue culture presents a reduced experimental variation due to a strict variable control and facilitates experimental replication. Although animal dissections are required to acquire primary tissue, the time and animal treatment stress associated with in vivo work is eliminated by in vitro models. When the material has passed in vitro tests then it can then be used in the fabrication of microsensors for biomedical applications. Later, these devices are evaluated in animals and, finally, they can be used in clinical studies. In this study, to evaluate the biocompatibility in vitro at the primary culture of dorsal root ganglia neurons (DRG) and electrophysiological recording of such cell in culture. As a biological system, we used DRG neurons, because we believe that when the neurons survive and have electrical activity in cultured materials, this is a clear indication of biocompatibility.

The purpose of this study is to evaluate the electrical properties of a-$Si_{1-x}Ge_x$:H films deposited by LF-PECVD and to study its biocompatibility. For the study of biocompatibility, we decided to use as a first approach the use of dorsal root ganglion (DRG) neurons obtained from rats. These neurons are relatively easy to obtain, and its electrophysiological behavior has been widely characterized, thus providing an optimal system in order to evaluate cell survival and functionality. This work is organized as follows: in Section 2, the deposition properties, electrical characterization a-$Si_{1-x}Ge_x$:H films, and the DRG cell culture and electrophysiological recordings on surface a-$Si_{1-x}Ge_x$:H films are described. The results of the electrical performance characteristics and results of biocompatibility are described in Section 3. Finally, we present the conclusions and future developments of research in Section 4.

2. Methodology

Generally, the study of biocompatible materials in electrophysiology use gold, titanium, parylene or platinum-black microelectrodes [21]. However, these materials have electrical limitations such as very high impedance and oxidation in electrolyte environment. Recently, some alternative materials have been studied to improve the electrical characteristics between the electrolytic environment and microelectrodes [21]. The a-$Si_{1-x}Ge_x$:H films were fabricated on a Corning 2947 glass substrate deposited by a LF-PECVD plasma system (planar reactor Reinberg, model AMP3300) with the following conditions: Glow discharge was maintained for one hour at 110 kHz with a substrate temperature of 300 °C, a low pressure (8.10^{-4}), and a power of 600 Watts. The feed gases were Silane (SiH_4) and Germane (GeH_4), and the dilution gas was hydrogen (H_2). The SiH_4 flow was kept constant and undiluted. In addition, the GeH_4 flow was diluted at 10% in H_2. The germanium concentration in the gas phase is defined as:

$$X_{Ge} = \frac{Q_{Ge}H_4}{\left(Q_{Ge}H_4 + Q_{Si}H_4\right)} \tag{1}$$

$QGeH_4$ and $QSiH_4$ are the gas flows of germane and silane, respectively. Four processes were made with a flow variation of GeH_4, $0 \leq XGe \leq 1$. Table 1 shows the gas flows in standard cubic centimetres per minute (sccm) of SiH_4, GeH_4, H_2, and the germanium concentration (XGe).

Table 1. Gas flow for the deposition of a-$Si_{1-x}Ge_x$:H thin films

Sample	X_{Ge}	$QGeH_4$ (sccm)	$QSiH_4$ (sccm)	QH_2 (sccm)
991	1	500	0	1000
990	0.83	500	100	1000
989	0.63	150	100	1000
988	0	0	100	1000

In this study, experimental procedures involving animals were reviewed and approved by the Institutional Animal Care and Use Committe (CICUAL) from the Vice-Rector for Research and Graduate Studies (VIEP-BUAP). Animal care and experimental procedures were performed according to the Reglamento de la Ley General de Salud en Materia de Investigación para la Salud of the Secretaría de Salud de México (NOM-062-ZOO-1999). These procedures were made to minimize animal suffering and to reduce the number of animals used, as outlined in the ''Guide for the Care and Use of Laboratory Animals" that is issued by the National Academy of Sciences. For the cell culture, we used Wistar rats of postanatal days 7-10 (P7-P10) that were euthanized by decapitation [21]. The dorsal root ganglia were isolated and were incubated in culture medium L-15 (Gibco) added with 0.125% collagenase and 0.125% trypsin for 30 min at 37 °C. The dorsal root ganglia were subsequently washed with L-15 medium, and the cells dissociated by mechanical agitation. Dissociated cells were placed in a surface of a-$Si_{1-x}Ge_x$:H films and time was allowed for cells to adhere to the surface. Then, was added fetal bovine serum (FBS) 10% (Gibco) and L-15 medium modified for CO_2, supplemented with $NaHCO_3$ (10 mM) and Hepes (10 mM). The medium was complemented with penicillin 100 IU/ml (Merck). The cells were incubated at 37 °C in an atmosphere of air (95%) and CO_2 (5%), the CO_2 concentration was controlled by the incubator. The temperature was kept constant until the electrophysiological recording (18 to 24 hours). Individual neurons of the dorsal root ganglia were identified under a phase contrast invertoscope (Nikon TMS, Tokio, Japan). For electrical recording of the cultured neurons, glass capillary electrodes filled with an electrolytic solution were used. A typical patch clamp recording in the whole cell condition was used [22]. Recording pipettes were pulled from borosilicate glass capillaries using a Flaming- Brown (80/PC; Sutter Instruments, San Rafael, CA); they typically had a resistance of 1 to 3.5 MΩ when filled with the internal solution (Table II) and the seal resistance to the cell exceeded 1 GΩ. In this condition, setup allowed for recording ionic currents flowing through the cell membrane (voltage clamp condition), or the voltage responses of the cell produced by current injection (current clamp condition). For both types of recordings, we used a voltage Axopatch 200B clamp amplifier (Molecular Devices, Sunnyvale, CA). Command signals and data acquisition were generated by a 16 bit analog-digital converter (Digidata 1440, Molecular Devices) controlled by pClamp 10.0 software

(Molecular Devices). Voltage clamp and current clamp experiments were performed. In all experiments, once the whole-cell configuration was obtained, the cell capacitance and series resistance were electronically compensated (80%). Signals were low-pass filtered at 5 or 2 kHz and digitized at 20 or 10 kHz. Cells were continuously bathed with external solution (Table 2). Also, the internal solutions contained 2 mM ATP- Mg and 1 mM GTP-Na. The external pH solution was adjusted at 7.4 with NaOH and the internal pH solution was adjusted at 7.2 with KOH. Osmolarity was monitored by a vapor pressure osmometer (Wescor, Logan, UT) and was around 300 mOsm for internal solutions and for external solutions was adjusted with dextrose at 310 mOsm.

Table 2. Standard solution composition (in mM)

Solution	NaCl	KCl	$CaCl_2$	$MgCl_2$	HEPES	EGTA
External	140	5.4	1.8	1.2	10	-
Internal	10	135	0.134	5	5	10

3. Results and discussion

The Ambios technology atomic force microscope (AFM) in a non-contact mode was used to characterize the roughness of the surface of a-$Si_{1-x}Ge_x$:H films deposited at 300 °C by LF-PECVD on a Corning 2947 glass substrate. Three scales were used to register the roughness on the thin film surfaces; the first a 1x1 µm, second a 2.5 x 2.5 µm and finally 501.8 x 501.8 nm. AFM results are shown in Figure 1(a-c), the a-$Si_{1-x}Ge_x$:H H thin films exhibited finer surface roughness values in the nanometer range (11-15.4 nm).

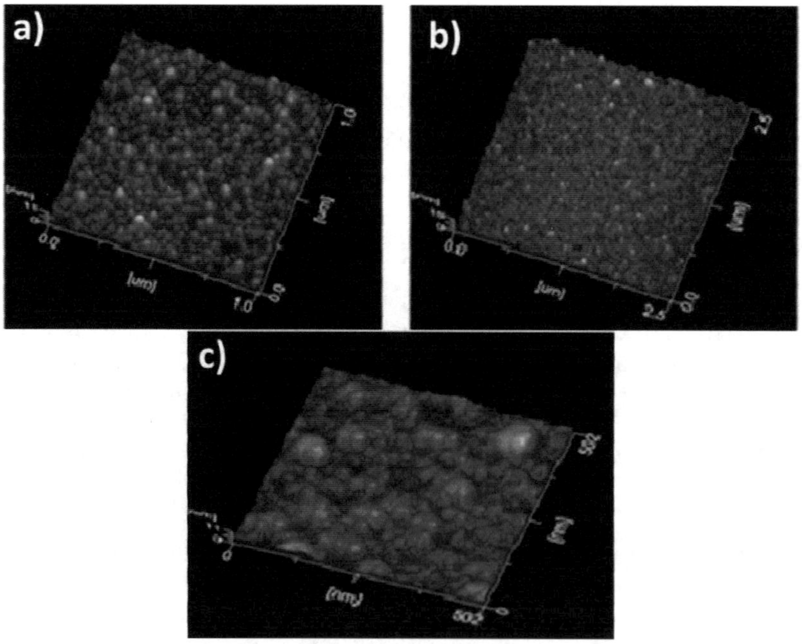

Figure 1. Surface morphology and roughness for a-$Si_{1-x}Ge_x$:H films

Current-voltage (I-V) curves for the a-$Si_{1-x}Ge_x$:H films samples with different germane compositions were plotted, the changes on the slope and the current between samples is caused by the variation of electrical resistance in every case (Figure 2). The electrical resistance decreased in the range of 50 MΩ to 4 kΩ as the amount of GeH_4 present in the samples increased.

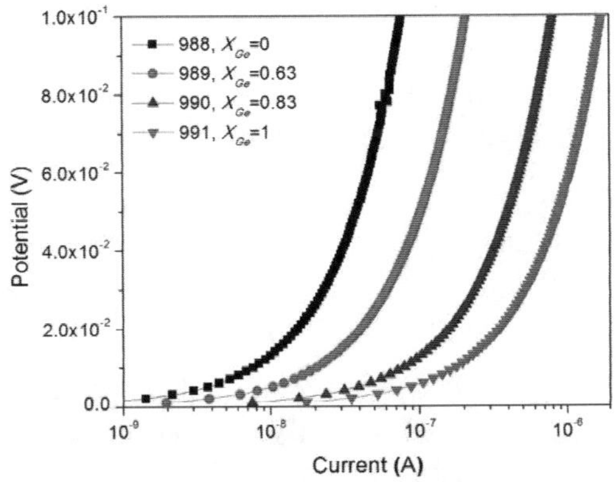

Figure 2. I-V curves of a-$Si_{1-x}Ge_x$ films deposited by LF-PECVD

We used the cells after 18 hours of incubation, because according to standard techniques previously described, it is enough time for neurons to express the ionic channels and membrane receptors after dissection and isolation. Additionally, with this culture technique we get better viability conditions without myelin on the cellular body. The DRG neurons grown on the a-$Si_{1-x}Ge_x$:H film surface were subject to electrophysiological recording to determine whether or not they were alive and functional, which was judged by their capability of generating membrane ionic currents and action potentials. Voltage clamp recordings showed that these neurons display typical ionic currents, including a fast- inward current at the beginning of voltage clamp pulse (most probably a Na^+ current) and ensuing outward currents (corresponding to a set of K^+ currents), as shown in Figure 3(a). In current clamp experiments, the membrane voltage of the cells was about -60 mV and under depolarizing current pulse injection the neurons were capable of firing typical action potentials similar to those previously described in standard substrate culture, as shown in Figure 3 (b).

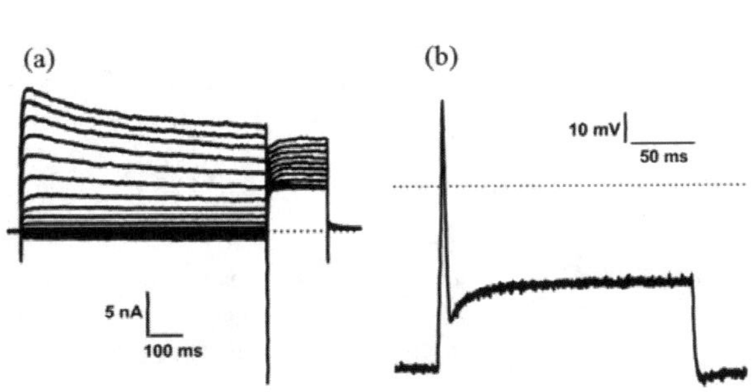

Figure 3. Recording of ionic currents in DRG neurons cultured in a-$Si_{1-x}Ge_x$:H films surface. (a) Total ionic current generated by a protocol of increasing voltage pulses from -110 to 70 mV starting from a holding potential of -60 mV. Recording of ionic currents shows that cells express the standard set of current found in vivo. (b) Action potential discharge produced by current pulse injection of 1 nA and 200 ms of duration. The dotted line indicates the zero current and voltage respectively.

4. Conclusions

The a-$Si_{1-x}Ge_x$:H thin films deposited by LF-PECVD at 300 °C temperature was studied, analyzing their morphology and biocompatibility. These thin films exhibited finer surface roughness values in the nanometer range suitable for biomedical applications. The experiments carried out confirmed the feasibility of using a-$Si_{1-x}Ge_x$:H thin films as a biocompatible material for the stimulation and recording of the electrical activity of DRG neurons in culture. The a-$Si_{1-x}Ge_x$:H thin films have surface property characteristics that could create an environment adequate for the cells. The electrophysiological recordings show that isolated DRG neurons obtained from the rat after 18 hours in incubation on a-$Si_{1-x}Ge_x$:H films surface displayed typical electrical behaviour, observed in standard conditions. The resistance of a-$Si_{1-x}Ge_x$:H thin films can be modified with different compositions of germane. Future work will include the deposit of these films on the surface microelectrodes with the readout circuit in the same chip under CMOS standard process.

5. References

[1] S. Dueñas, H. Castán, H. García, and L. Bailón, "Electrical characterization of High-K dielectric material", Dielectric Material, InTech, Rijeka, pp. 213-250, 2012.

[2] R. Olivares, J. J. Olaya, C. Ramírez, and E. S. Rodil, "Biocompatibility of niobium coatings", Coatings, vol. 1, pp. 72-87, 2011.

[3] F. López-Huerta, B. Cervantes, O. González, J. Hernández-Torres, L. García-González, R. Vega, A. L. Herrera-May, and E. Soto, "Biocompatibility and Surface Properties of TiO2 Thin Films Deposited by DC Magnetron Sputtering", Materials, vol. 7, pp. 4105-4117, 2014.

[4] A. L. Herrera-May, J. C. Soler-Balcazar, H. Vázquez-Leal, J. Martínez-Castillo, M. O. Vigueras-Zuñiga, and
L. A. Aguilera-Cortés, "Recent Advances of MEMS Resonators for Lorentz Force Based Magnetic Field Sensors: Design, Applications and Challenges", Sensors, vol. 16, 1359, 2016.

[5] M. Galindo-Mentle, F. López-Huerta, R. Palomino-Merino, C. Zúñiga-Islas, W. Calleja-Arriaga and A. L. Herrera-May, "Fabrication process of a microstructures based on hydrogenated amorphous SiGe films for applications in MEMS devices", J. Mech. Sci, Technol., vol. 29, pp. 1673-1679, 2015.

[6] P. Gonzalez, B. Guo, M. Rakowski, K. De Meyer, and A. Witvrouw, "CMOS compatible polycrystalline silicon–germanium based pressure sensors", Sensor Actuat. A, vol. 188, pp. 9-18, 2012.

[7] F. López-Huerta, M. Galindo-Mentle, C. Zúñiga-Islas, W. Calleja-Arriaga, R. Palomino-Merino, J. De la Hidalga, and V. Castaño, "Boron-doped hydrogenated amorphous semiconductor MEMS". Mat. and Tech., vol. 49, pp. 3-8, 2015.

[8] J. El-Rifair, S. Sedky, R. V. Hoof, S. Severi, D. Lin, S. Sangameswaran, R. Puers, C. Van Hoof, and A. Witvrouw, "SiGe MEMS at processing temperatures below 250 °C", Sensor. Actuat. A, vol. 188, pp. 230-239, 2012.

[9] A. D. Sundararajan, and S. M. Rezaul Hasan, "CMOS integrated elliptic capacitive pressure sensor in SiGe MEMS". Microsys. Technol., vol. 20, pp. 145-155, 2014.

[10] A. Khaled, M. Raoof, V. Cherman, K. Jans, M. Abbas, Sh. Ebrahim, G. Bryce, P. Verheyen, A. Witvrouw, and I. De Wolf, "Effect of the functionalization process on the performance of SiGe MEM resonators used for bio-molecular sensing", Microelectron. Reliab., vol. 52, pp. 2272-2277, 2012.

[11] P. González, B. Guo, M. Rakowski, K. De Meyer, and A. Witvrouw, "CMOS compatible polycrystalline silicon-germanium based pressure sensors", Sensor. Actuat. A, vol. 188, pp. 9-18, 2012.

[12] M. Makihata, S. Tanaka, M. Muroyama, S. Matsuzaki, H. Yamada, T. Nakayama, U. Yamaguchi, K. Mima,
Y. Nonomura, M. Fujiyoshi, and M. Esashi, "Integration and packaging technology of MEMS-on-CMOS capacitive tactile sensor for robot application using thick BCB isolation layer and backside-grooved electrical connection", Sensor. Actuat. A, vol. 188, pp. 103-110, 2012.

[13] B. Bouxin, K. Maier, A. Hackner, G. Mueller, F. Shao, J. D. Prades, F. Hernandez-Ramirez, and J. Ramon Morante, "On-chip fabrication of surface ionisation gas sensors", Sensor. Actuat. B, vol. 182, pp. 25-30, 2013.

[14] C. I. Chang, M. H. Tsai, Y. C. Liu, C. M. Sun, and W. Fang, "Pick-and-place process for sensitivity improvement of the capacitive type CMOS MEMS 2-axis tilt sensor", J. Micromech. Microeng., vol. 23, 095029, 2013.

[15] P. V. Cicek, Q. Zhang, T. Saha, S. Mahdavi, K. Allidina, F. Nabki, and M. E. Gamal, "A novel prototyping method for die-level monolithic integration of MEMS above-IC", J. Micromech. Microeng., vol, 23, 065013, 2013.

[16] W. Hernández-Montero, E. Zaldívar-Huerta, C. Zúñiga-Islas, A. Torres-Jacome, C. Reyes-Betanzo, and A. Itzmoyotl-Toxqui, "Optical and compositional properties of amorphous silicon-germanium films by plasma processing for integrated photonics", Opt. Mat. Exp., vol. 2, pp. 358-370, 2012.

[17] J. Cho, D. P. Pham, J. Jung, C. Shin, J. Park, S. Kim, A. H. Tuan Le, H. Park, S. M. Iftiquar, and J. Yi, "Improvement of hydrogenated amorphous silicon germanium thin film solar cells by different p-type contact layer", Mater. Sci. Semicond. Process., vol. 41, pp. 480-484, 2016.

[18] S. D. Kuhn, K. Sridharan, Z. Hao, P. Muir, M. Suresh, A. Singh, and V. S. Raj, "Biocompatibility of uncoated and diamond like carbon coated Ti–20%Hf alloy", Mater. Sci. Technol., vol. 24, pp. 575-578, 2013.

[19] B. Cervantes, F. López-Huerta, R. Vega, J. Hernández-Torres, L. García-González, E. Salceda, Emilio, A.L. Herrera-May, and E. Soto, "Cytotoxicity Evaluation of Anatase and Rutile TiO2 Thin Films on CHO-K1 Cells in Vitro", Materials, vol. 9, pp. 1-11, 2016.

[20] E. Soto, A. Limón, A. Ortega, and R. Vega, "Características morfológicas y electrofisiológicas de las neuronas del ganglio vestibular en cultivo", Gac. Med. Mex., vol. 138, pp. 1-13, 2002.

[21] E. Salceda, A. Garateix, A. Aneiros, H. Salazar, O. López, and E. Soto, "Effects of APC a sea anemone toxin on sodium currents of mammalian neurons", Brain Res., vol. 1110, pp. 136-143, 2006.

[22] N. Fertig, R. H. Blick, and J. C. Behrends "Whole Cell Patch Clamp Recording Performed on a Planar Glass Chip", Biophys. J., vol. 82, pp. 3056–3062, 2002.

Acknowledgements
The work was supported by CONACYT through grant 48757 and PRODEP 511-6/18-9245-UV-PTC-883.

IOP Conf. Series: Materials Science and Engineering **628** (2019) 012004 doi:10.1088/1757-899X/628/1/012004

Modified $BaTiO_3$/PVDF/Ni composites with high dielectric constant and low dielectric loss

Lin Gao[1], Zhengguang Sun[2*], Junlong Yao[2,3*], Feng Lu[3]

[1] School of Chemistry and Environmental Engineering, Jianghan University, Wuhan 430056, China
[2]Ministry-of-Education Key Laboratory for the Green Preparation and Application of Functional Materials, Hubei University, Wuhan 430062, China
[3]School of Materials Science and Engineering, Wuhan Institute of Technology, Wuhan 430073, China

Corresponding author: sunshine@hubu.edu.cn; junlongyao@yahoo.com

Abstract. Low dielectric properties of BaTiO3/PVDF composites limit its application in modern microelectronics and high energy storage. In this work, three-phase composites with high dielectric properties were prepared based on the surface modification of BaTiO3(BT) ceramic. The results show that, after modification of BT, the dielectric constant of the BT/PVDF composite increased by 44%, and the dielectric loss decreased by 20% at 100Hz. Nickel powder was added to the modified composites for increasing dielectric constant, when the addition of nickel powder was 20%, the dielectric constant of composites was up to 441 and the dielectric loss of composite was only 0.057.

1. Introduction

High dielectric-constant materials with low dielectric loss are widely used in the field of the capacitor [1]. BT/PVDF composites show relatively high dielectric properties and good processability, which are potentially ideal dielectric materials. However, the composites can not replace the traditional inorganic dielectric material in the use of capacitor manufacture because of the relatively low dielectric constant[2-4].

A series of methods have been taken to improve the dielectric properties of BT/PVDF composites [5] [6]. Surface modification for BT, such as surface hydroxylation, silane coupling agent modification, can enhance the dielectric properties of BT/PVDF composites, which owing to the better interface connection within composites[7-8]. But there are still some problems, such as the high dielectric loss.

In this work, BT was modified by tartaric acid, and a series of composites were fabricated. It was found that the dielectric properties of BT/PVDF composites were improved after modification. Furthermore, three-phase composites of PVDF, which are modified BT and nickel power, exhibit higher comprehensive dielectric properties.

2. Experimental

2.1 Material

Barium titanate was purchased from Nantong the new electronics co., LTD., China Nickel powder (high purity) and Tartaric acid（AR）was obtained from Tianjin days reagent co., LTD., China.

Polyvinylidene fluoride was supplied by Shanghai institute of organic fluorine, China. All chemicals were used as received.

2.2 Preparation of composites

The surface modification of BT involved several steps: (i) dissolving tartaric acid in 50ml distilled water and the solution was adjusted to pH 9, (ii) dispersing BT in the solution and put into three-necked flask, (iii) heating to 50°C and stirring for 2 hr under water bath, (iv) drying at 60 °C overnight under blast oven.A series of composites were prepared by following a simple blending and hot-pressing at about 170-175 °C and 10-12MPa for 30 min (Figure 1). Finally, samples were polished and slivered by sliver-paint.

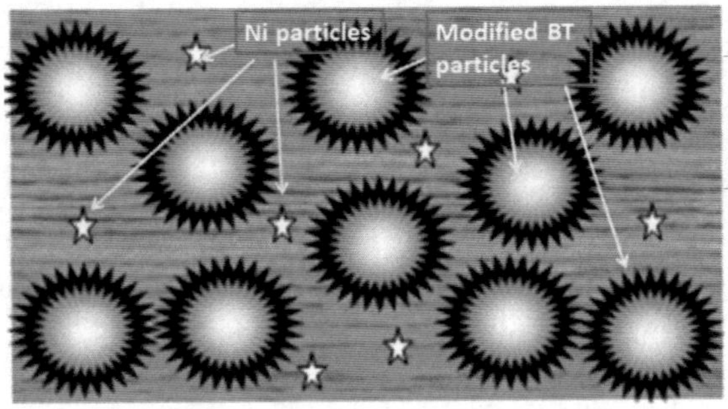

Figure 1. Modified BT/PVDF/Ni composite

2.3 Characterization

The measurement of dielectric constant and electric conductivity were carried out using a HIOKI3532-50 LCR HITESTER meter (HIOKI, Japan) in the frequency range from 100 Hz to 2MHz. The dielectric constant was calculated by using the equation (1).

$$\varepsilon = Cd/\varepsilon_0 S \quad (1)$$

C is the capacitance (farads), ε0 the free space dielectric constant value (8.854×10-12F/m), S the capacitor area (m2) and d the thickness (m) of the samples.Scanning electron microscope (SEM) is from a HitachiS-530 (JEOL Japan) scanning electron microscope.

3. Results and discussion

It can be seen in Figure 2(a) and (b) that the dielectric constant of BT/PVDF composites increase with the increase of BT volume fraction(f_{BT}), and the composites show low dielectric loss. The dielectric constant of the composite material increases from 121 to 174 Figure 2(a), an increase of 44%, while the dielectric loss of the system increased from 0.051 to 0.042 in Figure 2(b), decreased 18% after modification.

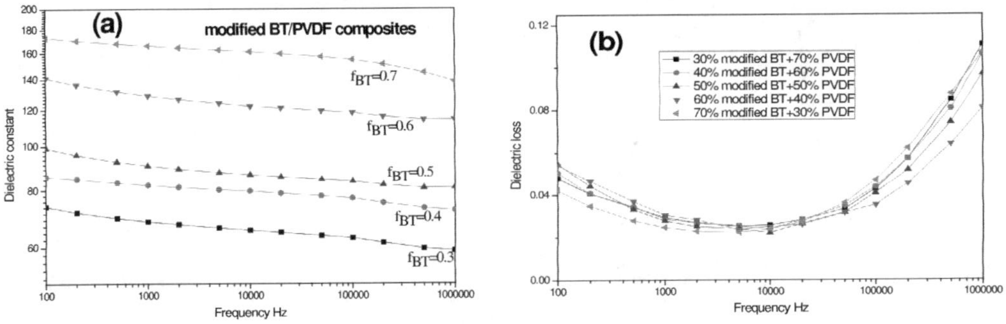

Figure2. The dielectric properties of modified BT/PVDF composites: (a), the dielectric constant of composites ;(b), the dielectric loss of composites

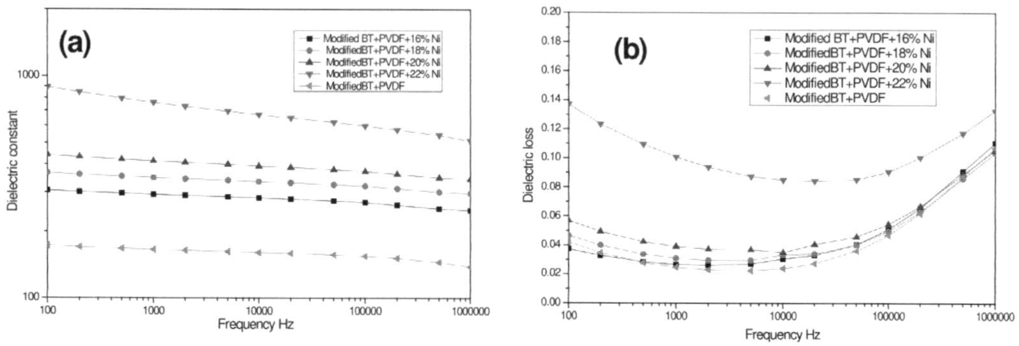

Figure 3. The dielectric constant (a) and loss (b) of modified and unmodified BT/PVDF/Ni composites

Nickel power is introduced into the modified BT/PVDF composites to further increase the dielectric constant owing to the percolation theory. It can be seen in Figure 3, that the addition of Nickel power can significantly improve the dielectric constant of the composites (volume fraction PVDF keep 50%). When the volume fraction of Ni arrive 22 %, the system reaches a maximum dielectric constant, 894 at 100Hz Figure 3(a), when the volume faction of nickel power is 20%, the dielectric loss is only 0.057 Figure 3(b), and the dielectric constant approaches to 441, increase of 153% comparing with modified BT/PVDF composites.

Figure 4. SEM of modified and unmodified BT/PVDF/Ni composites: A, B unmodified BT/PVDF/Ni composite; and C, D modified BT/PVDF/Ni composite

SEM images of cross section of the BT/PVDF/Ni composites with 20%Ni are shown in Figure 4. The spherical particles are given to nickel power, and the lump solid for BT, the floc is for PVDF. It can be clearly found that the dispersibility of BT in the composites is good, and the surface of BT particles becomes much rougher after modifying, even there can be some thrusting.
The reason of emergence of this thrusting is that an interface layer will be formed if BT particle modified by tartaric acid, which can enhance the interface connection of the composites. Furthermore, comparing to Ni dispersion before and after modification, it can be known that the agglomeration of Ni is reduced. The percolation threshold vale increase a lot with the reduction of agglomeration. It can be known that modification can increase the dielectric constant of the composites.

4. Conclusion

Modified BT/PVDF composites which have higher dielectric constant and lower dielectric loss were prepared. When the volume fraction of modified BT was 70%, dielectric constant was increased from 121 to 174, up 44%, dielectric loss was decreased from 0.05 to 0.04 at 100Hz. Modified BT/PVDF/Ni composites were fabricated, higher dielectric constant of 441 is observed at 100Hz with 20 vol% of Nickel powder, increased by 153% comparing with to modified BT/PVDF composites, and the dielectric loss is only 0.057; Modified BT/PVDF/Ni composites show excellent comprehensive dielectric properties due to good dispersibility of BT and Ni particles, and interface reaction mechanism will be studied in further.

5. References

[1] Jiongxin Lu, C Wong. 2008 *IEEE Transactions on Dielectrics and Electrical Insulation,* **15** 1322
[2] Shao J, Zhao Y M ,Yang C M, 2009 *High Voltage Apparatus* **45** 155
[3] Jung-Rag Yoon, Jeong-Woo Han, Kyung-Min Lee. 2009 *Transactions On Electrical And Electronic Materials* **10** 116
[4] Peter B, Shiva B, Yogesh A, 2009 *Materials* **2** 1697
[5] Yong Li, Xingyi Huang, Zhiwei Hu, 2011 *Applied Materials and Interfaces* **3** 4396
[6] Zhimin. D, Yang S, Cewen. N, 2002 *Applied Physics Letters* **81** 4813
[7] ZhiMin D, Haiyan W, Haiping X, 2006 *Applied Physics Letters* **89** 1
[8] Tao Z, Junwei Z, Ruiyao C, 2011 *Appl. Mater. Interfaces* **3**2184

Acknowledgments

The authors gratefully acknowledge the financial supports from the Open Project Fund of Ministry-of-Education Key Laboratory for the Green Preparation and Application of Functional Materials.

IOP Conf. Series: Materials Science and Engineering **628** (2019) 012005 doi:10.1088/1757-899X/628/1/012005

Study of nano bioactive glass for use as bone biomaterial comparison with micro bioactive glass behaviour

N Rocton, H Oudadesse, S Mosbahi, L Bunetel, P Pellen-Mussi and B Lefeuvre

University Rennes, CNRS, ISCR-UMR 6226, F-35000, Rennes, France

Corresponding author's e-mail: hassane.oudadesse@univ-rennes1.fr

Abstract. This research is based on the study of bioactivity kinetic in function of the glass particles size. Bioactive glasses have been elaborated in the ternary system SiO2-CaO-P2O5. Nano bioactive glass and micro bioactive glass have been synthesized by using two different processes. They are destined for use as bone biomaterials. The comparison was focused on the kinetic of the development of a calcium phosphate layer on their surfaces after immersion in a Simulated Body Fluid (SBF). The first bioactive glass BG is a melting-made glass with a particles size of about 60 μm. The second bioactive glass NBG is a sol-gel made glass through an emulsion system of synthesized particles of about 110 nm. The growing of the calcium phosphate layer at the surface of the glasses has been followed using several physicochemical techniques. Obtained results show the development of a calcium phosphate layer similar to carbonated hydroxyapatite. It crystallises in a hexagonal system with an P63/m space group. While melting-made glass needs 14 days to develop carbonated hydroxyapatite like crystal, sol-gel needs only 3 days to develop similar crystals. This difference offers wide opportunities and complementarities for the use of nano or micro bioactive glasses in the biomedical field.

1. Introduction

Hench invented bioactive glasses [1-3] in the Na2O-CaO-P2O5-SiO2 system, their utility has been widely proven in the area of bone restoration. Synthetic biomaterials are a good option for replacing biological bone grafts. Bioactive glasses, calcium phosphate and calcium carbonate are examples of synthetic materials used as biomaterials in orthopaedic surgery. They have the advantage of avoiding troubles linked to disease transmission. Thus, they are available in unlimited quantity, and they can be fitted for any person contrary to heterografts and xenografts. Precise studies to evaluate the biocompatibility and the physicochemical properties are required before the introduction of new synthetic materials for use in the biotechnological field. This work is focused on bioactive glasses. These glasses are particularly interesting due to their capacity to form a carbonated hydroxyapatite layer (HA) Ca10(PO4)6(OH)2. HA layer is very similar to the crystal composing the human bones.

Bioactive glasses may be synthesised using two different processes, both presenting advantages. The first is the synthesis of glass by fusion, it is historically used in many works [4-8]. The second presents an alternative using the polycondensation of a silica network via a so-called sol-gel synthesis [9-14].

This study is focused on the system CaO-P2O5-SiO2 which have shown some potential in their properties on biomedical field [15-17]. The synthesis and characterization of two bioactive glasses will be developed using the two different methods mentioned earlier. The development will focus on the effect of bioactive particles size on the kinetic of bioactivity.

In this study, nano-sized and micro-sized bioactive particles in the system CaO-P2O5-SiO2 were synthesized. Their chemical composition, microstructure, and bioactivity were investigated. Obtained results allow for a wider range of applications, particularly
in the field of biotechnology including but not limited to bone restoration,
bones regrowth, anti-bacteriological or metallic prothesis coating.

2. Materials and methods

2.1 Glass synthesis

2.1.1 Micro bioactive *glass (BG) synthesis process.* Preparation of the bioactive glass (46S6) by melting was previously described in our works [18]. The 46S6 synthesized in this study is a quaternary glass with the following molar percentage composition: 46 SiO2, 24 CaO, and 6 P2O5 and 24 Na2O. The melting process for the elaboration of the bioactive glass use silicon oxide (SiO2), Sodium metasilicate (Na2SiO3), Sodium metaphosphate (Na3P3O9), Calcium metasilicate (CaSiO3). These elements were weighed then mixed in a polyethene container for 2 hours in a planetary mixer. Origin and purity of these chemicals can be seen in table 1.

Table 1. Origin and purity of products for the melting process.

Chemical Name	Source	Mole fraction purity [a]
Silicon oxide (SiO_2)	Sigma-Aldrich Chemistry	0.9980
Sodium metasilicate (Na_2SiO_3)	Fisher chemical	$\geq$ 0.97
Sodium metaphosphate ($Na_3P_3O_9$)	Alfa Aesar	0.95
Calcium metasilicate ($CaSiO_3$)	Alfa Aesar	$\geq$ 0.97

[a] Purity of these elements has been controlled but has not been purified again.

Premixed materials were melted in a platinum crucible placed in an electric furnace. The first rise of temperature rate was 10 °C min-1 from 20 °C to 900 °C for a duration of 1 hour to decarbonate all products previously prepared, it is the calcination phase. After this first step, a new rise of temperature at 20 °C min-1 was conducted. The temperature was stabilized to 1350 °C for 3 additional hours, it is the melting phase. The thermal elaboration process is described in figure 1. After this 3-hour period of melting, the sample was air quenched and transferred to a preheated brass mould. The last step, the glass has been annealed at Tg ± 20 K (540 °C for 46S6) for 4 hours to stabilize the glass and to reduce physical tensions due to a brutal decrease in temperature during quenching. After this thermic treatment, the glass will be crushed to micrometres then subdued through a sieve to recover 60 ± 15 µm glass powder.

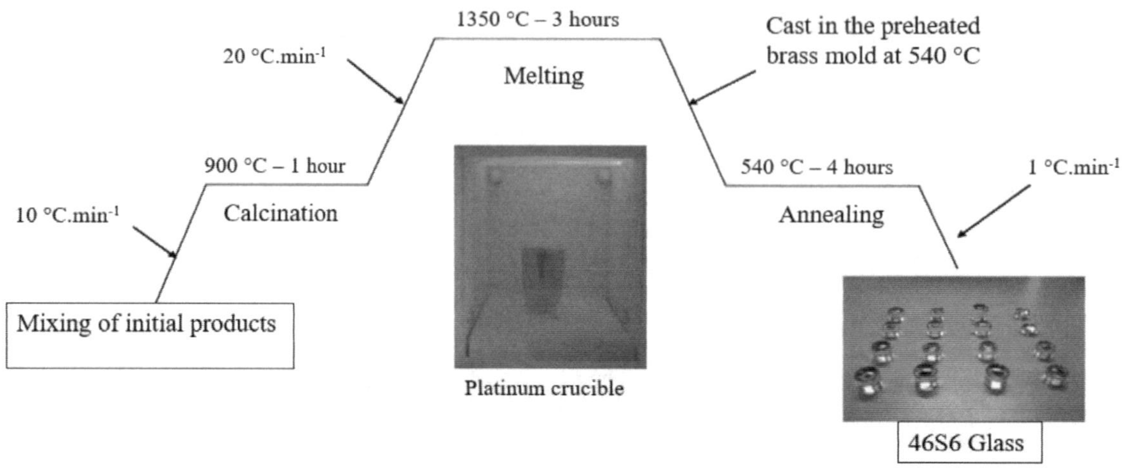

Figure 1. Thermal elaboration process for melting-made glass.

2.1.2 Nano bioactive *glass (NBG) synthesis process.* The nano bioactive glass (NBG) synthesized in this study is a ternary NBG with the following molar percentage composition: 55 SiO_2, 40 CaO, and 5 P_2O_5. The material was prepared from tetraethyl orthosilicate (TEOS), sodium trimetaphosphate ($Na_3P_3O_9$), calcium nitrate tetrahydrate ($Ca(NO_3)_2$ $4H_2O$) and ammonia (NH_3 H_2O). Furthermore, Octanol, Triton® X-100, and cyclohexane were also used as a surfactant and oil phase to prepare the emulsion. In this water in oil emulsion, the aqueous phase is dispersed to form microdroplets protect by a monolayer of surfactant molecules. The origin and purity of the product can be seen in table 2.

Table 2. Origin and purity of products for the sol-gel process.

Chemical Name	Source	Mole fraction purity [a]
Tetraethyl Orthosilicate (TEOS: $SiC_8H_{20}O_4$)	Merck	0.98
Sodium trimetaphosphate ($Na_3P_3O_9$)	Alfa Aesar	≥ 0.95
Calcium nitrate tetrahydrate ($Ca_2N_2O_6$ $4H_2O$)	Merck	0.99
Ammonia (NH_3 H_2O)	Carlo Erba Reagents	≥ 0.95
Octanol	Alfa Aesar	0.99
Triton® X100	Alfa Aesar	0.99
Cyclohexane	Fisher Chemical	≥ 0.95

[a] Purity of these elements has been controlled but has not been purified again

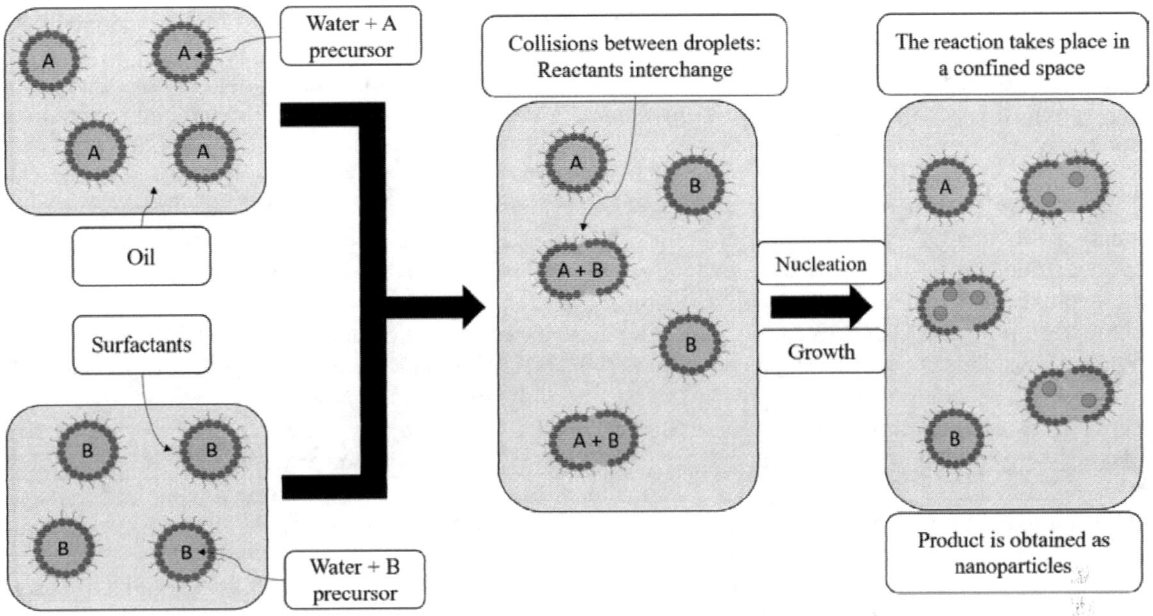

Figure 2. Diagram of the synthesis system for emulsion.

This process is based on the preparation and mixing of two different emulsions A and B.
Emulsion A is an aqueous ammonia phase containing TEOS and sodium trimetaphosphate. To this emulsion will be added the two surfactants (Triton® X-100 and octanol) and the oil phase (cyclohexane). This mixture will remain strongly stirred for one hour to allow the hydrolyzation of TEOS and sodium trimetaphosphate. Emulsion B is an aqueous phase based on calcium nitrate tetrahydrate, to which will also be added the same two surfactants and cyclohexane. This synthesis system is presented in a diagram in figure 2. Both emulsion compositions are resumed in table 3.

Table 3. Composition of each emulsions.

	Emulsion A	Emulsion B
Aqueous phase	Ammonia, water	Calcium nitrate tetrahydrate, water
Surfactants	Triton® X-100, Octanol	Triton® X-100, Octanol
Oil phase	Cyclohexane	Cyclohexane

After their separated preparation, the two emulsions were mixed together for a duration of 1 hour under a constant and vigorous speed then stirring have been stopped and the preparation let to rest for 45 minutes.
After this growing time, a white product was collected by centrifugal (6000 g) and washed repeatedly with ethanol to eliminate the most surfactant. Finally, drying to solid at 100 °C for 4 h and calcinated at 620 °C for 4 h to eliminate all surfactant trace.

2.2 Physicochemical characterization techniques

2.2.1 Size and *specific area measurement.* The determination of the size distribution of the synthesized nano particles was obtained using Dynamic Light Scattering (DLS) through Zetasier (Nano ZS, Malvern). This technique was used to measure the hydrodynamic diameter of the nanoparticles from 1 nm to 10 μm with ± 0.4 nm using the diffraction of light. Theses hydrodynamic measurements were performed at a 90 ° angle, after the dilution of the sample (50 μg mL^{-1}) in a saline solution to maintain

Brownian motion. The particles were dispersed in the saline solution for 15 min in an ultrasound bath to disperse possibly-formed agglomerates.

The specific surface area was calculated by the Brunauer-Emmett-Teller (BET) N_2 adsorption-desorption method (Flowsorb II 2300, Coultronics France SA), the total pore volume and the average pore diameter are also determined.

2.2.2 Thermal analysis. Thermal characteristics of the elaborated glasses (nano and micro) have been measured by using Differential Scanning Calorimetry (DSC). Measurements of thermal characteristics were carried out by means of a Setaram Labsys 1600TG-DTA/DSC thermal analyser under N_2 gas atmosphere at 0.1 MPa without pressure control. This analyser has been calibrated using a 3-peak calibration with $SrCO_3$, $BaCO_3$, and K_2CrO_4. First experimentation has been made using the same platinum crucible of 100 µL as the one used for all the analyses to eliminate the effect of the crucible in results. The microbalance sensitivity of the thermogravimetric simultaneous thermal analyser is better than 0.1 µg and its temperature precision is ± 1 K. DSC signal is used in this work. Bioactive glasses were studied at a heating rate of 5 °C min^{-1} raised from room temperature (around 20-25 °C) to 1350 °C. Each time, 40 ± 0.1 mg of glass powder was heated in a platinum crucible and at the same time, another empty platinum crucible was used as control (differential measurement).

2.2.3 Structure and functional groupings in elaborated materials. X-ray Diffraction (XRD) and Fourier transformed Infra-Red spectroscopy (FT-IR) have been used to control physicochemical properties of elaborated glasses. The control of the amorphous nature and the potential formation of new functional groups in these compounds were the main objectives of this characterization.

XRD patterns were recorded between 5 and 90 ° (2θ) by a PANalytical X'Pert PRO diffractometer with a precision of 0.026 ° by steps and a count time of 40 s per step using Cu Kα. Voltage was of 50 kV and the current was 40 mA during analyses. Powdered samples were prepared for XRD analysis and back-loaded into steel sample holders.

For FT-IR analysis, samples were embedded in KBr pellets and recorded by Bruker Equinox 55 spectrometer between 4000 cm^{-1} and 400 cm^{-1} in transmittance mode with a resolution of 2 cm^{-1}. FT-IR and XRD analyses were carried out at ambient temperature.

2.3 Soaking preparation for in vitro chemical reactivity for bioactive glasses

The in vitro bioactivity was evaluated by the immersion of 30 mg of nanopowder and 30 mg of micropowder in 60 ml of SBF [19] in the same conditions. The chemical composition of SBF is presented in table 4 and similar to the chemical composition of human blood plasma. These tests have been carried out for different duration of soaking composed by short (2, 4, 8, 16 h, 1, 3 days) or long delay (2 weeks, 1, and 2 months) and maintained at 37 °C under constant stirring. After soaking, samples have been removed from the SBF solution, rinsed with water and ethanol and dried at 75 °C for 3 h. The structure of each sample was analysed by XRD.

Table 4. Ionic composition of SBF compare to human blood plasma.

Ions (mM)	Na^+	K^+	Mg_2^+	Ca_2^+	Cl^-	HCO_3^-	HPO_4^{2-}	SO_4^{2-}
SBF	142.0	5.0	1.5	2.5	147.8	4.2	1.0	0.5
Plasma	142.0	5.0	1.5	2.5	103.0	27.0	1.0	0.5

Experimental concentration uncertainties u(C) = ± 0.1 mM.

2.4 Cytotoxicity tests

Cytotoxicity tests have been made for both glasses. The cytotoxicity of the micrometric glass has been studied and described in further research [20]. Due to the unknown harmfulness of nanoparticle powder, these tests have been made in a diluted solution of tween 20® added to a solution of Dulbecco's Modified Eagles Medium (DMEM). The DMEM is a solution used for the supply of nutrients necessary for the maintenance and proliferation of different types of cells in vitro. Cytotoxicity tests have been made for two different cell line: SaOs-2 (osteoblastic) and EA.hy926 (endothelial), two durations: 3 and 24 hours and three different concentration of nanoparticles: 25, 10, 1 µg/mL [21].

2.5 Statistical analysis.
For all the tests, data are expressed as the mean ± SD of three independent experiments.

3. Results and discussion

3.1 Physicochemical characterization of glasses

3.1.1 Size and specific area of melting and sol-gel bioactive glasses. Dynamic light scattering results show spherical particles with medium size around 110 nm for the sol-gel-made glass as shown in figure 3.

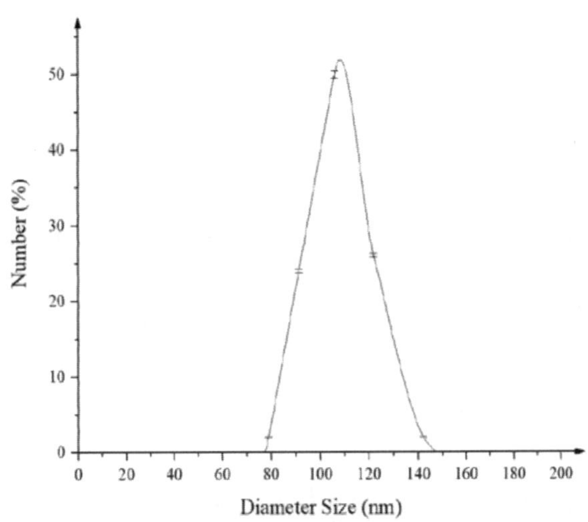

Figure 3. Size distribution of the nano bioactive glass using DLS.

The size of the glass particles synthesized by melting is ensured using sieves allowing the sizes to pass less than 40 µm and preventing the passage of sizes greater than 63 µm it results most of the particles are around 60 µm. BET obtained result show that the surface area decreases in function of particles size, from 69.4 to 1.1 $m^2\ g^{-1}$. These results can be seen in table 5

Table 5. Surface area, pore volume and pore diameter of studied glasses.

Glass size	Surface area ($m^2\ g^{-1}$)[a]	Pore volume ($cm^3\ g^{-1}$)[b]	Pore diameter (Å)[c]
110 nm	69.4	0.131	72.08
60 µm	1.1	0.001	33.16

[a] Experimental surface area uncertainty u(sa) = 0.1 m^2 g^{-1}.
[b] Experimental pore volume uncertainty u(pv) = 0.01 cm^3 g^{-1}.
[c] Experimental pore diameter uncertainty u(pd) = 0.01 Å.

A higher specific surface area enables a faster ion release for bioactive materials. This rise in ion release and ions transfer allow the possibility to obtain high healing kinetic for biomaterials.

3.1.2 Thermal *analysis results of studied bioactive glasses.* Each DSC spectra show the presence of a glass transition temperature, a crystallisation temperature and a melting temperature respectively called as Tg, Tc and Tf. According to literature [22], each synthetized material may be a glass. DSC spectra of materials are presented in figure 4 and 5.

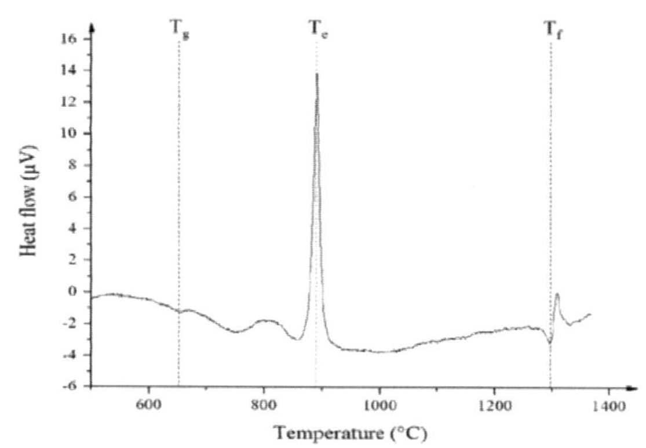

Figure 4. DSC of the sol-gel-made nano glass.

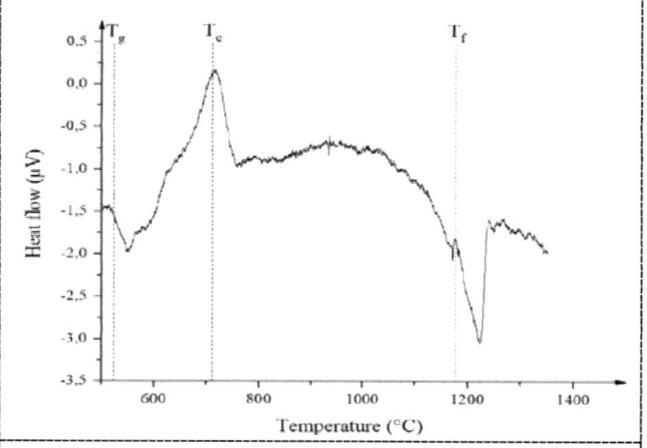

Figure 5. DSC of the melting-made micro glass.

3.1.3 Structure *and functional grouping of bioactives glasses after soaking.* FTIR results shown similar results for both glasses. The vibration strips at 470 cm^{-1} and 1200 cm^{-1} correspond to characteristic Si-O-Si bonds. While the vibration strips at 1400 cm^{-1}, 1070 cm^{-1}, 925 cm^{-1} and 567 cm^{-1} correspond to the phosphate and carbonate groups. These results are shown in figure 6.

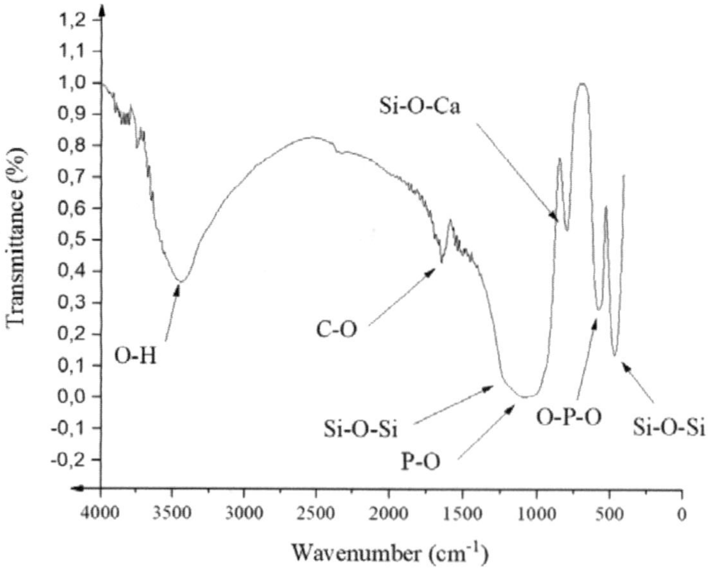

Figure 6. FTIR results of soaked glasses.

The XRD diffractogram of the micrometric material showed at figure 7 presents no sign of crystallisation before immersion. This diffractogram determine the micromaterial as glass. Results show a crystallisation of the glass into a similar form as the HA (use here as reference). The obtained results presented in this work show the first diffractograms with a crystallization effect. In the case of micrometric glass, it has been found that it is recognizable from the second week.

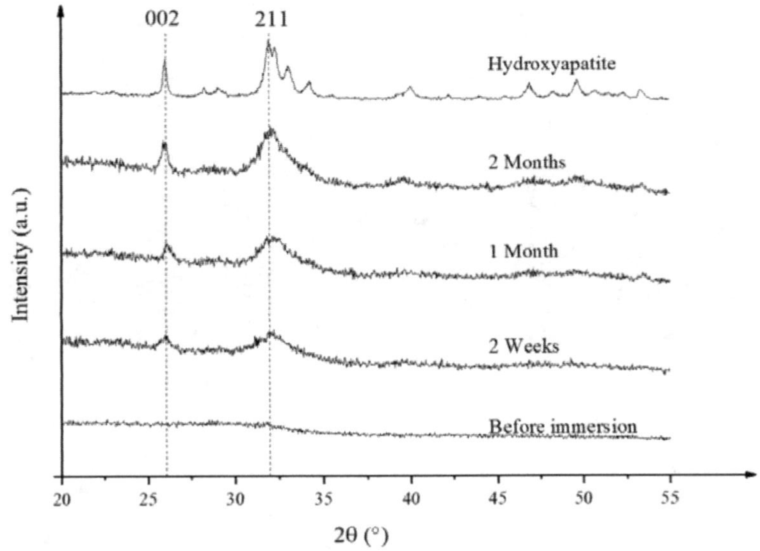

Figure 7. XRD diffractogram of melting-made micro glass before and after soaking.

The XRD diffractogram of the nanometric material showed at figure 8 doesn't presents sign of crystallisation before immersion. This diffractogram determine the nanomaterial as glass. Results show a crystallisation of the glass into a similar form as the HA. In the case of nanometric glass, it has been found that it is recognizable from the eighth hour.

The comparison between the two diffractograms allow to link the size of the bioactive particle with the specific surface area and the kinetic of crystallisation. In this case, the kinetics of crystallization is 4 times faster in the case of nanoscale bioactive glass when in that of micrometric glass whereas a factor of 70 separates their specific surface.

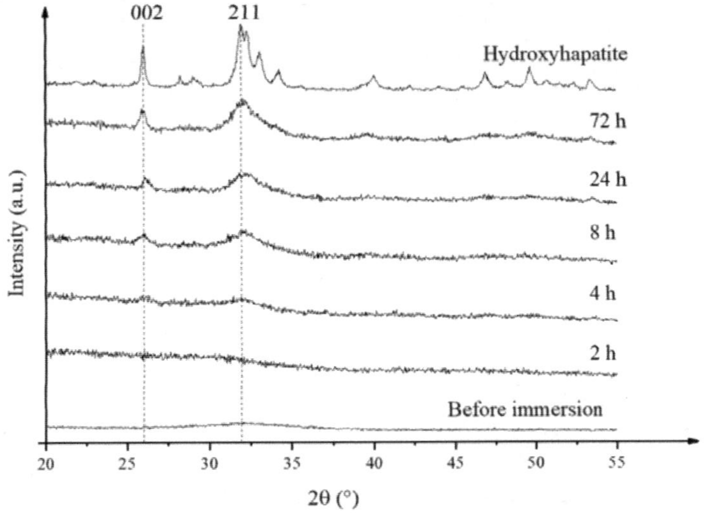

Figure 8. XRD diffractogram of sol-gel-made nano glass before and after soaking.

3.2 Cytotoxicity tests results

The cytotoxicity tests presented in figure 9 show that nanometric glass (110 nm) is relatively well tolerated by cells. These concentrations therefore have no toxicity for the cells studied. Nevertheless, the nanometric particles exhibit a phenomenon of aggregation once immersed in tween 20. The results therefore have great variation and are very difficult to reproduce as shown by the variations shown in the figure.

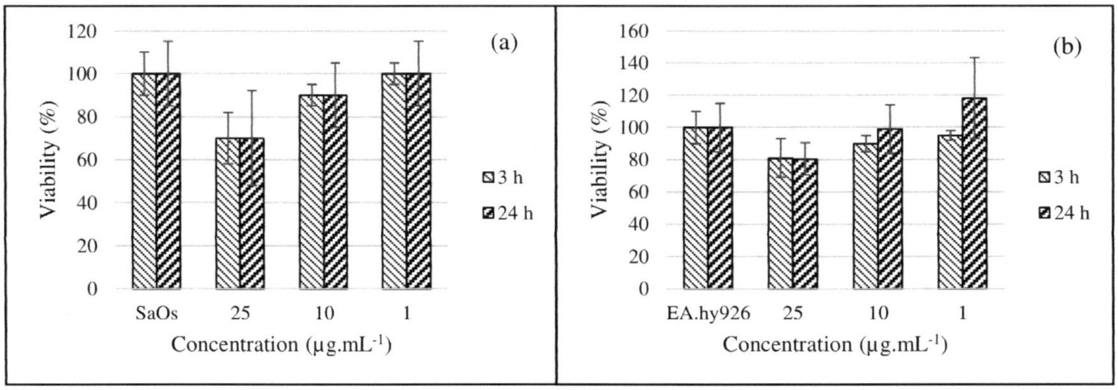

Figure 9. Viability of SaOs-2 (a) and EA.hy926 (b) cell lines as a function of nanoparticle concentration.

4. Conclusion

Despite major differences in the synthesis of the two materials presented in this article, the physicochemical characterization shows that they are bioactive glasses. Their characterisation highlights their amorphous nature and their capacity to form chemical bond and turn into crystal.
This study shows that changing the size of biomaterials causes changes in its specific surface area, pore size and ion exchanges capacity with its environment. In our case the reduction of a size ratio of 545 increases the specific surface area from 1.1 to 69.4 $m^2\ g^{-1}$ and makes it possible to obtain a crystallization similar to 2 weeks of immersion in 3 days of immersion only for the case of nano bioactive glass of 110 nm.
Nevertheless, it is important to highlight other differences that may affect the kinetics of bioactivity of the glasses studied during this research. Indeed, the composition, therefore the chemical environment of the vitreous matrix plays an important role in the ion exchanges between the medium (here SBF) and the surface of the glass. Consequently, glasses properties and particularly their composition may lead to the change on the kinetic of bioactivity.
In the end, the comprehension of all the elements and more particularly of the interactions of these elements between them is very important. Accurate knowledge will allow the development of biomaterials, effective, adaptable depending on the nature of the lesion undergone, its location and the constitution of the patient. With the aim of deepening the interaction of the size of the bioactive glass nanotubes and the kinetics of bioactivity, a study has been initiated and in progress to have more data to yield significant statistics.

5. References

[1] Hench LL, Splinter RJ, Allen WC and Greenlee TK, 1971 *J. Biomed. Mater. Res.* **5** 117
[2] Hench LL, 1998 *Biomaterials* **19** 1419
[3] Hench LL, 2002 *Science* **295** 1014
[4] Wers E, Oudadesse H, Lefeuvre B, Bureau B and Merdrignac-Conanec O, 2014 *Thermochimica Acta* **580** 79

[5] Wers E, Oudadesse H, Lefeuvre B, Lucas-Girot A, Rocherullé J and Lebullenger R 2014 *Journal of Thermal Analysis and Calorimetry* **117** 579

[6] Wers E, Oudadesse H, Lefeuvre B, Merdrignac-Conanec O and Barroug A 2015 *Applied Surface Science* **353** 200

[7] Franchini M, Lusvardi G, Malavasi G and Menabue L 2012 *Materials Science and Engineering: C* **32** 1401

[8] Dietrich E, Oudadesse H, Lucas-Girot A, Le Gal Y, Jeanne S and Cathelineau G 2008 *Applied Surface Science* **255** 391

[9] Zheng K and Boccaccini AR 2017 *Advances in Colloid and Interface Science* **249** 363

[10] Hench LL and West JK 1990 *Chemical Reviews* **90** 33

[11] Itoh H and Sugimoto T 2003 *Journal of Colloid and Interface Science* **265** 283

[12] Lukowiak A, Lao J, Lacroix J and Nedelec JM. 2013 *Chemical Communications* **49** 6620

[13] Nabian N, Jahanshahi M and Rabiee SM 2011 *Journal of Molecular Structure* **998** 37

[14] Saboori A, Rabiee M, Moztarzadeh F, Sheikhi M, Tahriri M and Karimi M 2009 *Materials Science and Engineering: C* **29** 335

[15] Hench L.L 1997 *Solid State Mater. Sci.* **2**, 604

[16] Itoh H and Sugimoto T 2003 *J. Colloid Interface Sci.* **265** 283

[17] Hench LL 2000 *Key Eng. Mater.* **192–195** 575

[18] Rocton N, Oudadesse H and Lefeuvre B 2018 *Thermochimica Acta* **668** 58

[19] Kokubo T, Kushitani H, Ohtsuki C, Sakka S and Yamamuro T 1992 *Journal of Materials Science: Materials in Medicine* **3** 79

[20] Oudadesse H, Dietrich E, Bui, XV, Le Gal Y, Pellen P and Cathelineau G 2011 *Applied Surface Science* **257** 8587

[21] Pellen-Mussi P, Tricot-Doleux S, Neaime C, Nerambourg N, Cabello-Hurtado F, Cordier S, Grasset F and Jeanne S *Journal of Nanoscience and Nanotechnology* 2018 3148

[22] J Zarzycki, *Les verres et l'état vitreux*, 1st ed., Masson, Paris, 1982.

IOP Conf. Series: Materials Science and Engineering **628** (2019) 012006 doi:10.1088/1757-899X/628/1/012006

Development of a PEMFC dynamic model and the application to the analysis of fuel cell vehicle performance

Wei Liu[1], Zhijun Peng[2], Bill Kim[3], Bo Gao[3], Yiqiang Pei[1]

[1]State Key Lab of Engines, Tianjin University, China
[2]Faculty of Creative Arts, Technologies and Science, University of Bedfordshire, UK
[3]AVL Powertrain, UK

Corresponding author's e-mail: Jun.Peng@beds.ac.uk

Abstract. In order to investigate basic output performances of PEMFC (Proton Exchange Membrane Fuel Cell) stack, a dynamic model of PEMFC stack has been developed by combining electrochemical sub-model and thermodynamic sub-model. With necessary validation, it demonstrates that modelling results and experimental data are in very good agreement in terms the U-I curve and power output. By applying the dynamic model to analyse performance outputs of PEMFC stack and applying the model for FC-Hybrid vehicle powertrain configuration, it demonstrates that improved PEMFC quality with increased maximum current density could increase the peak power output and also increase the working efficiency, although the increase of peak power is not linear relation with the increase of maximum current density. Higher working temperature of PEMFC would benefit the increases of both peak power output and efficiency. Compared to working temperature, ambient temperature's increase could also make positive influence on power output and efficiency, though the influence is weak. Coupling the dynamic model with a powertrain model of FC-Electric hybrid vehicle, the analysis suggests that both PEMFC stack and battery stack should have similar size for general driving condition. Too big either PEMFC stack or battery stack would increase the total weight then contaminate the fuel/energy economy.

1. Introduction

Along with the commercialisation of ground electric vehicles including hybrid vehicles and battery vehicles, the development of FCVs (Fuel Cell Vehicles) has been rapidly accelerated in recent years. Toyota Mirai, Honda Clarity, Hyundai ix35 PEMFC have been sold for several years. Compared to plug-in battery (hybrid or pure battery) electric vehicles, it is boosted that FCVs have obvious advantages for using more clean energy [1, 2].

Researches around fundamentals of fuel cells have been carried out widely and deeply in both academic and industrial fields, with both advanced experimental methods and numerical simulations. For meeting the product development of FCVs, more and more works for the FC powertrain configuration and the integration with other relating systems are urgently required, in terms of improved power density, cell durability, more effective catalyst loading, manufacturing cost reduction, and so on [3, 4]. While a comparable driving range as compared to HEVs (Hybrid Electric Vehicles, normally with combustion engines) and BEVs (Battery Electric Vehicles) is currently very important from the customer's viewpoint, fuel consumption or energy consumption is always the firstly critical factors for vehicle developers.

As there exist highly coupled, non-linear transport phenomena and electrochemistry in fuel cell, modelling therefore plays an important role in giving an understanding of what is happening inside a cell where the typical length scale of a few microns to millimetres poses some difficulties on experimental work. In addition to wide CFD (Computational Fluid Dynamic) investigations to internal flows and external cooling systems [5, 6], various control models [7, 8] and dynamic models [9, 10] have been developed and applied for studying some fuel cell components' performances or entire stack systems' operating performances.

It is therefore the objective of this study to build a dynamic model of PEMFC stack which can be easy to couple with FCV powertrain model for exploring the powertrain's working performances under various design and operating conditions.

2. Model description

2.1 Model structure and sus-models

The dynamic model of PEMFC stack mainly consists of two main sub-models, electrochemistry and thermodynamic models. As shown in Figure 1, the electrochemistry includes Nernst potential model, ohmic losses, concentration losses and activation losses. The thermodynamic model combines mass and energy conservation, heat transfer, water management, etc.

2.2 Model validation

To validate the dynamic model of PEMFC stack, experimental data based on a 2 kW Horizon PEMFC stack have been used. The test system is shown in Figure 2.

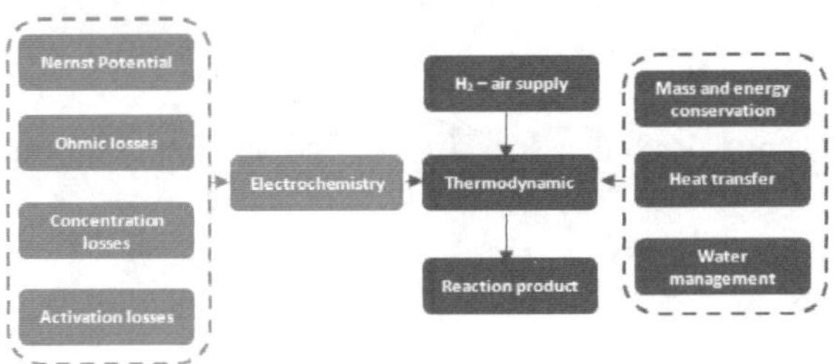

Figure 1 PEMFC dynamic model structure

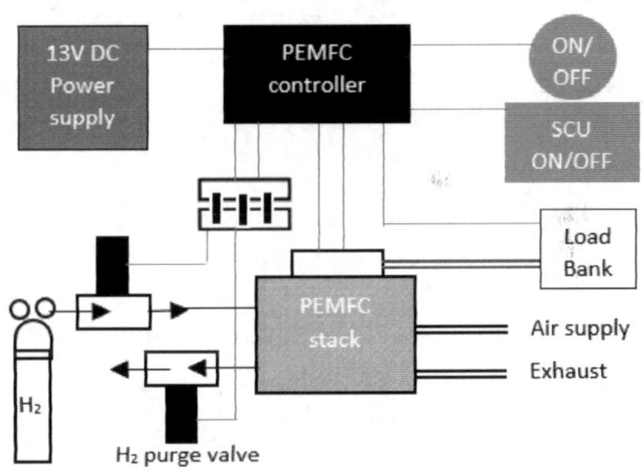

Figure 2 Fuel cell stack and test system

From the results shown in Figure 3, it demonstrates the modelling output for the U-I curve and power output as function of current. When the model can produce a very reliable results which are in very good agreement with the experimental data, as reflected on U-I curve and power output, it suggests the model can be used to simulate PEMFC output with adequate confidence.

3. Results and discussion

Based on successful validation to the dynamic model, PEMFC output performances and effects of some design parameters and operating parameters were studied with the model. In the final section of 'Results and Discussion', the PEMFC dynamic model is used for coupling with a FC-Electric hybrid vehicle model to examining the optimal integration between PEMFC stack and battery stack.

3.1 FC Stack performances

Applying PEMFC (Proton Exchange Membrane Fuel Cell) stack to vehicle powertrain, it is more concerned of the fuel consumption rate and powertrain energy efficiency. As shown in Figure 4, modelling results of SHC (Specific Hydrogen Consumption) and efficiency as function of power output of fuel cell stack are presented and compared with experimental results. Basically, the fuel cell stack's efficiency is following the variation of the voltage and with some similar trend. At the peak power point, the efficiency is about 40%, which is very similar as diesel engines' efficiency of peak power.

But the efficiency at part-load is much higher than full load, and it reaches the max value while the power output is going down to the point close the idle condition. Based on this, the specific hydrogen consumption is lower under low load condition than high load condition.

In terms of the comparison between experimental results and modelling results, it shows the model has produced very successful prediction for the efficiency. Although for the hydrogen consumption rate, there is some small gap between experimental data and modelling results at some power value, it has very good agreement for the trend.

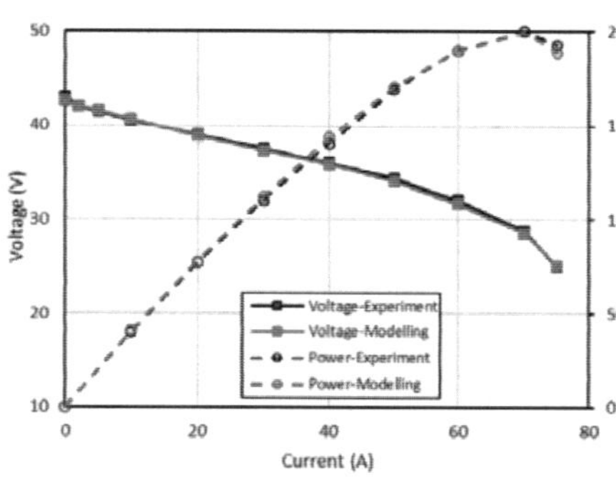

Figure 3 Model validation result

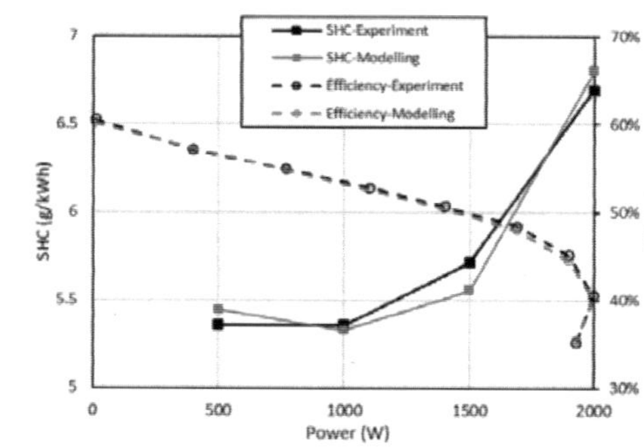

Figure 4 SHC (Specific Hydrogen Consumption) and efficiency as function of power

3.2 Effects of design parameters

For this case, three kinds of PEMFC with different maximum current densities 0.5 A/cm2, 1.0 A/cm2 and 1.5 A/cm2, respectively, are compared. As different current density reflects different design and manufacture quality, the higher the current density is, the higher the power output produces for the same fuel cell size.

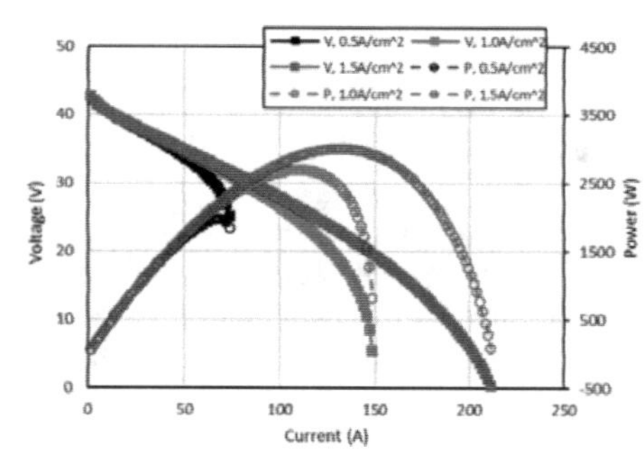

Figure 5 Effects of fuel cell quality on outputs of voltage and power

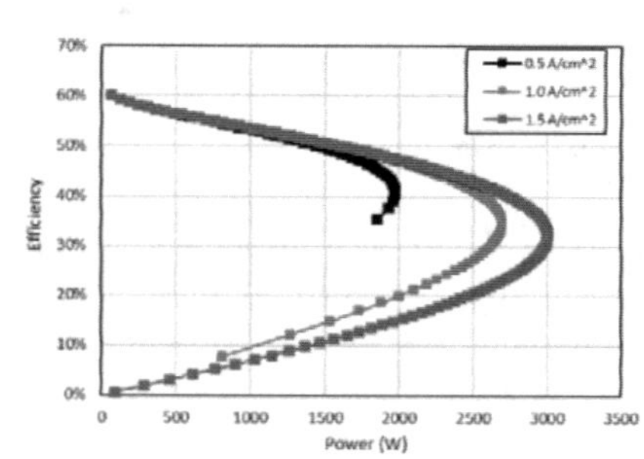

Figure 6 Variation of efficiency as function of power with different fuel cell quality

The voltage and power outputs as function of the current density are demonstrated in Figure 5. From U-I and Power-I curves, it can be found, although higher maximum current density can result in voltage output at higher current and make the maximum power output higher, the increase of power output is not linear as the increase of maximum current density. From 0.5 to 1.0 to 1.5 A/cm2 of maximum current density, the peak power outputs are 1.97 kW to 2.69 kW to 3.00 kW. The efficiencies for three different current density are presented in Figure 6. From the results, it suggests that efficiencies at the peak power point get apparent reduction with the increase of maximum current density. With 0.5 A/cm2 of maximum current density, the efficiency at the peak power of 1.97 kW is about 41.16%. But it decreases to 34.64% at 2.69 kW for 1.0 A/cm2, and to 32.02% at 3.00 kW for 1.5 A/cm2.

If keeping all of those three maximum current densities working at 41% of efficiency, the power output can increase from 1.97 kW to 2.50 kW and 2.67 kW, respectively.

3.3 Effects of operating parameters

To find how the working temperature influences the output performances, the simulation has been carried out with the dynamic model and results are presented in this section. As shown in Figure 7 and Figure 8, the results are produced by just changing the working temperature but keeping other parameters constant. From the U-I curve, it shows that under low current condition, the voltage output keeps decrease with the increase of working temperature. This is mainly due to increased working temperature making the activation losses increased. Under high current condition, higher working temperature makes the concentration losses reduced, resulting in higher voltage output. Then this is beneficial for some obvious increase of peak power output. For 30, 65 and 100 °C of working temperature, the peak power output can increase from 1.79 to 1.97 to 2.14 kW, respectively.

In addition to the increase of maximum power output by increased working temperature, the working efficiency has also apparent increased under high load condition. As presented in Figure 8, the efficiency at maximum power output point (1.79, 1.97 and 2.14 kW) are 38.31%, 41.16% and43.79%, respectively. If keeping the fuel cell stack working under 45% of efficiency, the power outputs are 1.64 kW, 1.87kW and 2.12 kW, respectively, for three working temperatures.

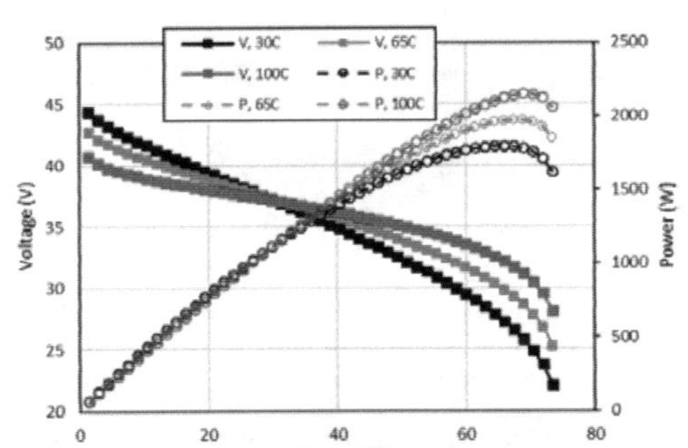

Figure 7 Effects of working temperature on outputs of voltage and power

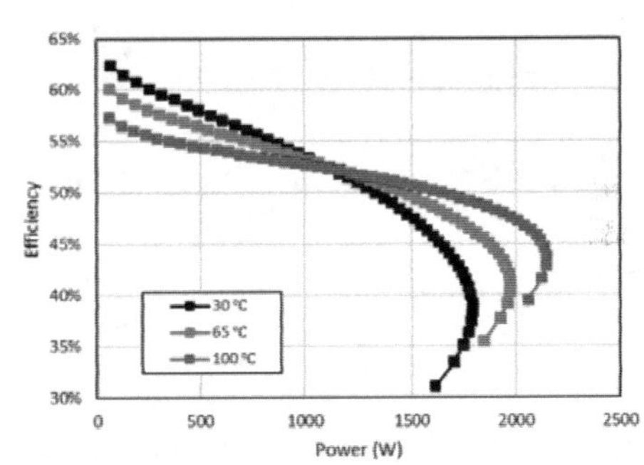

Figure 8 Effects of working temperature on efficiency

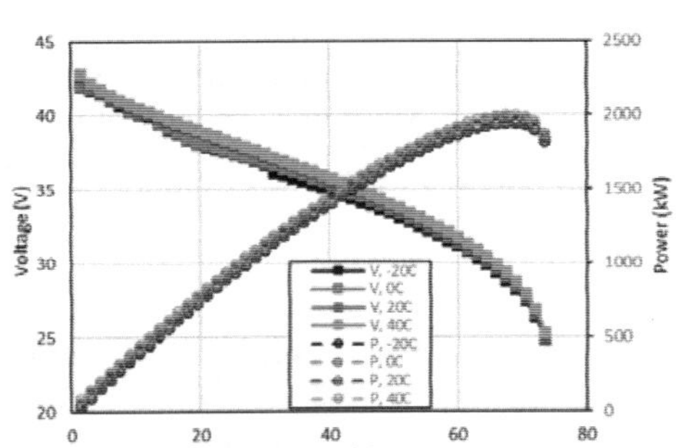

Figure 9 Effects of ambient temperature on outputs of voltage and power

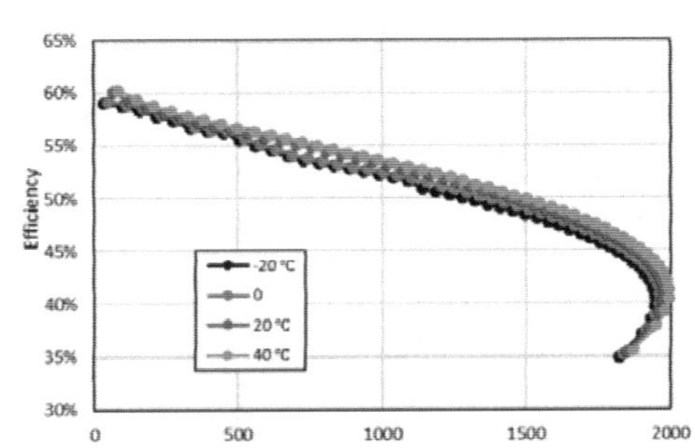

Figure 10 Effects of ambient temperature on efficiency

The next case is for changing ambient temperature but keeping other parameters constant. As shown in Figure 9, the voltage output will have slight increase with increased ambient temperature, this should be due to reduced ambient temperature reducing heat losses from the PEMFC stack to cooling air. Then from -20 to 40 °C, the maximum power output can increase from 1.94 kW to 1.99 kW, about 2.58% increase rate. As shown in Figure 10, effects of ambient temperature on PEMFC efficiency can't be also ignored, in particular under part-load conditions where are the general working area of road vehicles. For instance, if the ambient temperature increases from -20 °C to 40 °C at 1 kW, the efficiency can increase from 52.23% to 53.61%, with an increased rate of about 2.64%. At the peak power point, the efficiencies are 40.53% and 41.30% for -20 °C and 40 °C, respectively.

3.4 Application on vehicle performance analysis

By applying the dynamic model of PEMFC stack into vehicle powertrain model, different hybrid configurations (FC-Electric), as shown in Figure 11, were explored. The top configuration is more like current passenger FCV as developed in Toyota Mirai. The vehicle model can also consider other cases, such as plug-in hybrid (FC-Electric) as shown as the middle one in Figure 11, and much bigger battery stack as shown as the bottom one in Figure 11. By analysing the hybrid configuration detail with the vehicle powertrain model, coupled with the PEMFC dynamic model, the optimal integration between PEMFC stack and battery stack has been produced, as shown in Figure 12.

As shown in Figure 12, in order to achieve the lowest fuel/energy consumption, the optimal integration between PEMFC stack and battery stack is about 50 kW battery size and 60 kW PEMFC stack, for the analysed vehicle case. To increase either PEMFC stack size or battery stack size would make the powertrain too heavy, then increase fuel/energy consumption. If the battery pack is too small, the fuel/energy consumption would also increase because of the lack of adequate energy regeneration.

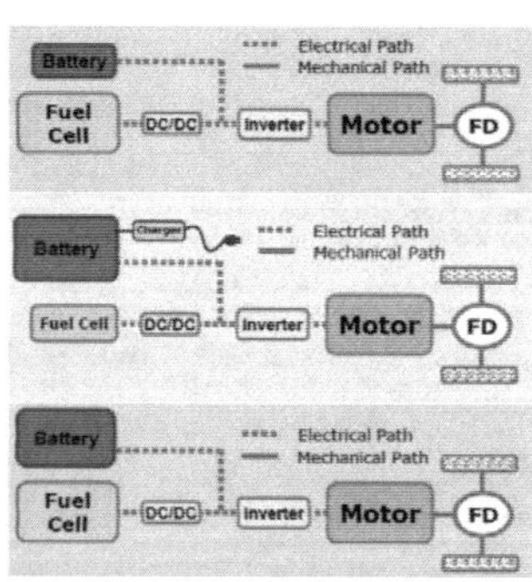

Figure 11 FCV powertrain configuration

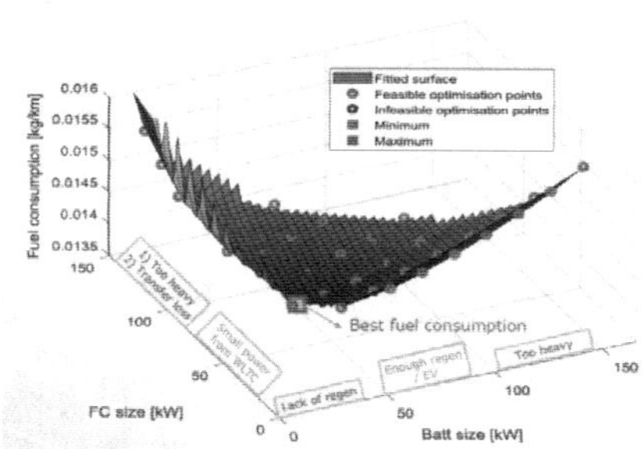

Figure 12 Optimised FC stack size and battery size for FCV

4. Conclusions

In this reported research, a dynamic model of PEMFC stack has been developed by combining electrochemical sub-model and thermodynamic sub-model. The validation demonstrates that modelling results and experimental data are in very good agreement in terms the U-I curve and power output. By applying the dynamic model to analyse performance outputs of PEMFC stack and applying the model for FC-Hybrid vehicle powertrain configuration, the following conclusions have been derived.

- Improved PEMFC quality with increased maximum current density could increase the peak power output and also increase the working efficiency, although the increase of peak power is not linear relation with the increase of maximum current density.
- Higher working temperature of PEMFC would benefit the increases of both peak power output and efficiency.
- Compared to working temperature, ambient temperature's increase could also make positive influence on power output and efficiency, though the influence is weak.

- Coupling the dynamic model with a powertrain model of FC-Electric hybrid vehicle, the analysis suggests that both PEMFC stack and battery stack should have similar size for general driving condition. Too big either PEMFC stack or battery stack would increase the total weight then contaminate the fuel/energy economy.

5. Recommendations for the future research – a). the model can be optimised to include different cooling conditions, such as air cooling and water cooling. b). the model can be improved to consider possible leaking of hydrogen, because it has been found hydrogen leaking is an important factor for influencing hydrogen consumption rate in practical PEMFC vehicles.

6. Reference

[1] Pouria Ahmadi, Seyed_Hosein Torabi, Hadi Afsaneh, Yousef Sadegheih, Hadi Ganjehsarabi, Mehdi Ashjaee, The effects of driving patterns and PEM fuel cell degradation on the lifecycle assessment of hydrogen fuel cell vehicles, International Journal of Hydrogen Energy, https://doi.org/10.1016/j.ijhydene.2019.01.165, 2019.

[2] Farah Ramadhani, M. A. Hussain, Hazlie Mokhlis, Muhamad Fazly, Jarinah Mohd Ali, Evaluation of solid oxide fuel cell based polygeneration system in residential areas integrating with electric charging and hydrogen fueling stations for vehicles, Applied Energy, Volume 238, Pages 1373-1388, 2019.

[3] Zhifeng Xia, Bowen Wang, Zirong Yang, Kangcheng Wu, Qing Du, Kui Jiao, Effect of operating conditions on performance of proton exchange membrane fuel cell with anode recirculation, Energy Procedia, Volume 158, Pages 1829-1834, 2019.

[4] Tao Chen, Shihua Liu, Jiwei Zhang, Mengnan Tang, Study on the characteristics of GDL with different PTFE content and its effect on the performance of PEMFC, International Journal of Heat and Mass Transfer, Volume 128, Pages 1168-1174, 2019.

[5] Guobin Zhang, Kui Jiao, Multi-phase models for water and thermal management of proton exchange membrane fuel cell: A review, Journal of Power Sources, Volume 391, Pages 120-133, 2018.

[6] Z. Peng, P. Choopanya, CFD Investigation into Internal Flows of PEM Fuel Cell for Optimal Performances, Advanced Materials Research, Vol.724-725, PP757-761, 2013.

[7] Zhixiang Liu, Lun Li, Yi Ding, Huiwen Deng, Weirong Chen, Modeling and control of an air supply system for a heavy duty PEMFC engine, International Journal of Hydrogen Energy, Volume 41, Issue 36, Pages 16230-16239, 2016.

[8] Lianghui Huang, Jian Chen, Zhiyang Liu, Mohamed Becherif, Adaptive thermal control for PEMFC systems with guaranteed performance, International Journal of Hydrogen Energy, Volume 43, Issue 25, Pages 11550-11558, 2018.

[9] A. Saadi, M. Becherif, D. Hissel, H. S. Ramadan, Dynamic modeling and experimental analysis of PEMFCs: A comparative study, International Journal of Hydrogen Energy, Volume 42, Issue 2, Pages 1544-1557, 2017.

[10] Zhen Wu, Peng Tan, Bin Chen, Weizi Cai, Meina Chen, Xiaoming Xu, Zaoxiao Zhang, Meng Ni, Dynamic modeling and operation strategy of an NG-fueled SOFC-WGS-TSA-PEMFC hybrid energy conversion system for fuel cell vehicle by using MATLAB/SIMULINK, Energy, Volume 175, Pages 567-579, 2019.

Acknowledgement

The authors would like to acknowledge the financial support to the work reported here by Innovate UK through the project of FCHCV (113618), and the contribution from research assistants Xinyu Wang, Zhisheng Gu and Keni Huang for result processing.

IOP Conf. Series: Materials Science and Engineering **628** (2019) 012007 doi:10.1088/1757-899X/628/1/012007

Effect of sintering parameters on the densification and hardness of NiAl-CNTs composite

M A Awotunde [1], O O Ayodele [1], A O Adegbenjo [1,3], A M Okoro [1], M B Shongwe [2], P A Olubambi [1]

[1] Center for Nanoengineering and Tribocorrosion, School of Mining, Metallurgy and Chemical Engineering, University of Johannesburg, South Africa
[2] Institute for NanoEngineering Research, Tshwane University of Technology, Pretoria, South Africa
[3]Mechanical Engineering Department, The Ibarapa Polytechnic, Eruwa, Oyo State 200005, Nigeria

corresponding author's e-mail : *mary.awotunde@uniben.edu

Abstract. Nickel aluminides have attracted considerable interest in the past few decades owing to its unique properties. In this work, nickel aluminide (NiAl) was formed in-situ during spark plasma sintering of admixed powders of nickel, aluminium and carbon nanotubes (CNTs) after ball milling. 1 wt % CNTs was incorporated into the intermetallic matrix and the effect of varying sintering parameters investigated, particularly the sintering temperature and pressure. Results showed that a combination of higher sintering temperature with lower pressure yielded better results than lower sintering temperature and higher pressure. Thus the former parameters yielded better densification and subsequently higher micro hardness values of the NiAl-CNTs composites as compared to the latter.

1. Introduction

Aluminides have shown huge potential in high temperature applications due to their high temperature strengths, lightweight, high wear resistance and high oxidation resistance [1-3]. In particular, interest in nickel aluminides have progressively grown over the years because of the aforementioned properties in combination with their relative lightweights as compared to expensive nickel base super alloys [4]. The major limiting factor of these aluminides is their room temperature brittleness due to insufficient slip systems within the lattice [5, 6]. Efforts have been made in the research community to eliminate this challenge mostly by introducing an additional element into the dual system to enhance the ductility and toughness [7]. Recently, carbon nanotubes (CNTs) have proven to be ideal reinforcing candidates owing to the enhanced mechanical properties exhibited by the CNT reinforced metal matrices. They have been widely incorporated into metallic matrices like copper, titanium and aluminium. The mechanical properties of metal matrices have been seen to depend on the microstructure [4]. The microstructures of these metals however, also depend on their process history. The most popular fabrication route for these aluminides have been the powder metallurgy route due to its ease of fabrication coupled with the advantage of producing near net shape components [8]. During sintering, the metallurgical interactions occurring within the matrix are responsible for the microstructure, and consequently the resulting mechanical properties. The fabrication parameters, especially sintering parameters play a critical role in determining the resulting mechanical properties and consequently, material performance. In this work, 1 wt % CNTs were incorporated into nickel aluminide intermetallic matrix via ball milling and the sintering parameters varied. The resulting mechanical properties from each combination of sintering parameters were thus investigated.

2. Methodology

2.1. Starting materials
Commercially available nickel, supplied by Weartech Limited with particle size of 0.5-3.0 microns and 99.5% purity, spherical aluminium powder with average particle size of 25μm was supplied by TLS Technik GmbH & Co., and MWCNTs supplied by Nanocyl, Belgium with average diameter of 9.5 nm and average length of 1.5μm were used in this study.

2.2. Low energy ball milling
The nickel, aluminium and CNTs powders were ball milled together to effectively disperse the CNTs in the nickel and aluminium powders using a planetary ball mill, PM 100 with milling speed of 150rpm. Nickel and aluminium powders were set at a weight ratio of 1:1 in all the sample mixtures for the synthesis of the NiAl composites. Ball to powder ratio of 10:1 was used and a 10 min break was taken after every 10 min milling to prevent excessive over heating of the powders. Weight percent of CNTs was set at 1 wt % for all samples for accurate comparison and evaluation. The powders were milled for 6 h after which the mixed powders were consolidated by spark plasma sintering.

2.3. Spark plasma sintering
The admixed powders were compressed in a graphite die and the whole die assemble was placed in the SPS system (model HHPD-25, FCT Germany). To investigate the effect of sintering parameters on the mechanical properties, 20 mm diameter with 5 mm height discs were sintered with two different sets of sintering parameters namely: 1) sintering temperature of 800 °C and pressure of 32 MPa and 2) sintering temperature of 600 °C and pressure of 50 MPa. Holding time was 5 min and heating rate was set at 100 °C per min for both regimes.

2.4. Density and hardness measurements
The densities of the sintered discs were measured using the Archimedes' principle according to ASTM standards. The recorded density was an average value of five repeated measurements taken from each sample and the relative densities calculated as percentages of the composites' theoretical densities. Vickers indentation microhardness tests were conducted on sectioned samples that had been previously polished to achieve mirror-like surfaces for all the samples. A load of 100 gf and 10 s dwell time was used for the hardness tests. The reported microhardness values are the arithmetic mean of five successive indentations made on the smooth sample surfaces.

3. Results and discussion
Figure 1 shows the powder morphologies of the starting powders that were milled together. This was done to effectively disperse the CNTs within the powders and secondly to trigger the reaction between nickel and aluminium to form an intermetallic powder matrix.
After 6 h of milling time, the CNTs clusters were effectively broken down and the CNTs uniformly distributed as observed in Figure 2. Low energy ball milling was used throughout to prevent significant damage of the CNTs in the matrix which would be detrimental to their mechanical properties [9]. This is because the mechanical properties of CNTs reinforced composites are largely dependent on the structural integrity of the CNTs. Hence the damage of these nanotubes would adversely affect the mechanical properties of the composites. Figure 3 shows the transmission electron microscopy (TEM)

images of the nanotubes, depicting the homogeneous dispersion of the nanotubes and thus endorsing Figure 2. This confirms that the ball milling process employed was particularly effective in dispersing intact nanotubes with well retained tubular morphologies.

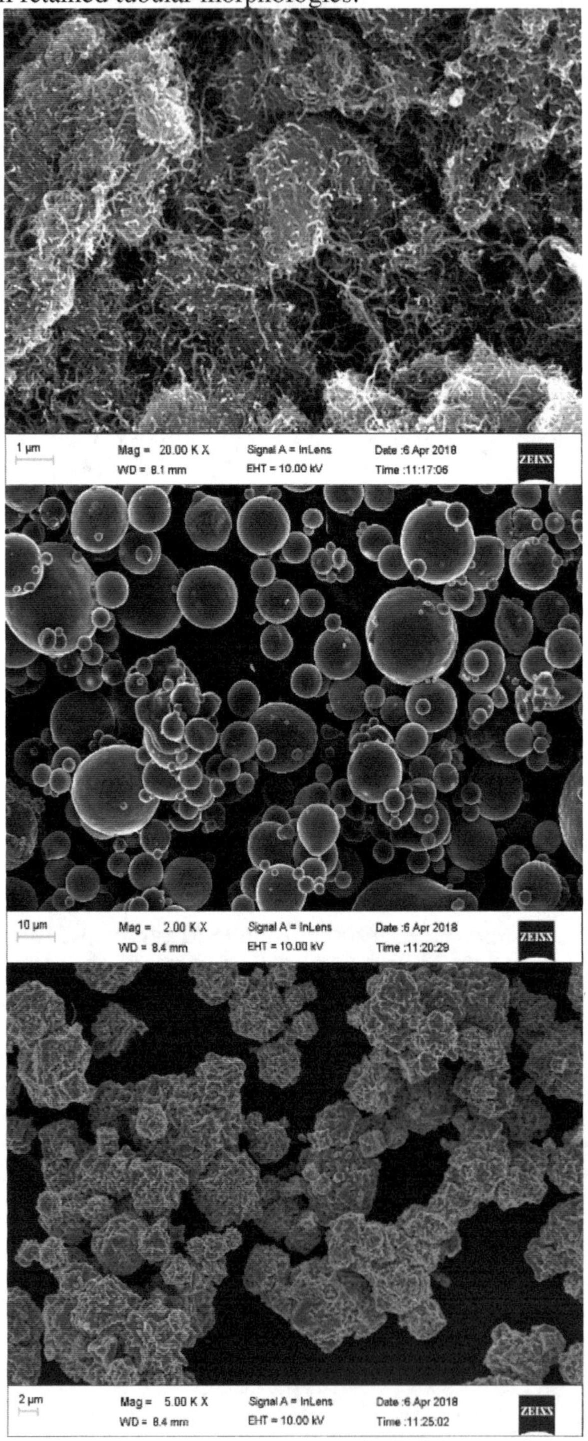

Figure 1. SEM images of starting powders (a) CNTs (b) aluminium and (c) nickel.

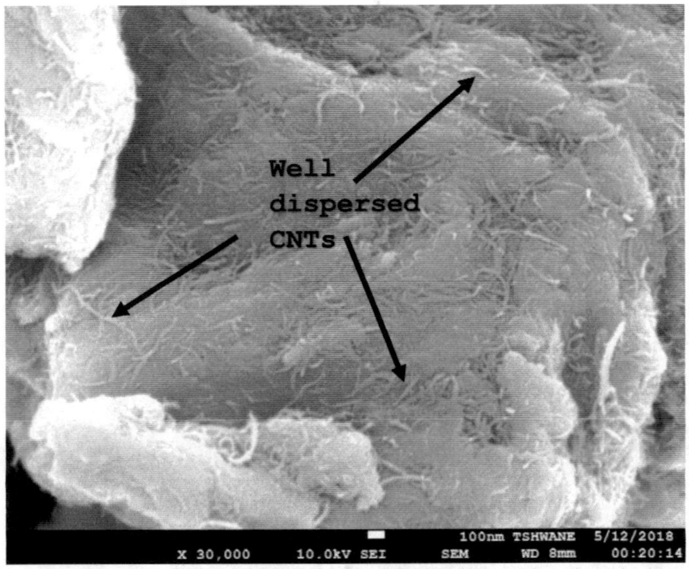

Figure 2. SEM image showing 1 wt % CNTs uniformly dispersed in nickel aluminide powders after 6h of low energy ball milling.

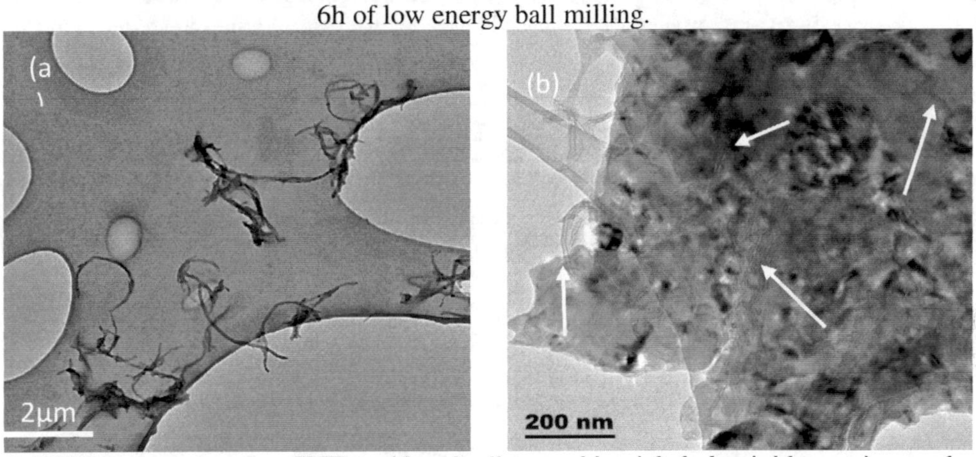

Figure 3. TEM images showing CNTs uniformly dispersed in nickel aluminide matrix powders after 6h of low energy ball milling.

3.1. Relative density and micro hardness

The correlation between sintering parameters and densification cannot be downplayed. Densification, to a large extent, is dependent on sintering parameters. Sintering at higher temperatures or higher pressures tend to give better densification during the sintering process. As can be observed in Table 1, the effect of sintering parameters on the relative density and micro hardness of the NiAl-1 wt % CNT composites is quite significant in this study.

Table 1. Effect of sintering parameters on the densification and hardness of NiAl-1wt% CNT composite

Sample ID	Sintering temperature (°C)	Sintering pressure (MPa)	Relative density (%)	Porosity (%)	Hardness (HV/GPa)
A	800	32	97.5	2.5	839/8.23
B	600	50	96	4	691/6.78

Both samples A and B have the same composition of starting powders of CNTs, nickel and aluminium with the same milling history. Sample A was sintered at a higher sintering temperature with reduced pressure, while sample B was sintered at a reduced temperature and higher sintering pressure. The relative densities of the consolidated composites showed a higher density for sample A that experienced a higher sintering temperature. The relative density was 1.5% higher than that of sample B which was sintered at a lower temperature. The observed reduction in the density and subsequently microhardness of sample B depicts the presence of microvoids and microholes within the microstructure [10]. The presence of these flaws makes the compact bulk more susceptible to failure as these defects are potential sites for crack formation. This behaviour is similar to that which was observed in Cooke et al, [11] who also observed higher densification with higher sintering temperature of 550 °C compared to the sample sintered at 400 °C. Chen et al, [12] similarly in his work observed an improvement in densification with increase in sintering temperature.

The seemingly insignificant 1.5% increase in densification led to a substantial increment of 148HV in microhardness of sample A, suggesting that densification played a crucial role in the resulting mechanical properties of sintered composites. The reason for this increment can be attributed to improved bonding between the powder particles at higher sintering temperatures favouring better interfacial contact [12] thus reducing the porosity, leading to better densification and ultimately higher hardness. Given that diffusion is a thermally activated process, at higher sintering temperatures, diffusion is higher, facilitating the rapid diffusion of atoms between the matrix and reinforcement, leading to a cleaner interface between them [13]. This tends to improve the interfacial bonding strength between the matrix and reinforcement, consequently enhancing the load transfer mechanism leading to higher strengthening effects in the composites, hence the improved mechanical properties observed in the composite sintered at 800 °C.

Additionally, the hardness value obtained in this study for Sample A is significantly higher (almost 100% higher) than that documented in literature for NiAl-CNT composites. Ameri et al, [14] in a similar study, achieved a relative density of 85.6 % and corresponding microhardness value of 4.16 GPa in the NiAl-1 % CNT composite. The density and microhardness values observed in this study were 97.5 % and 8.23 GPa respectively, even though the sintering temperature used in Ameri et al, [14] was significantly higher (1000 °C) than that employed in this study. Groven et al, [15] also obtained microhardness values of 4.5 GPa with the incorporation of single walled CNTs and a much lower value of 3.5 GPa with the incorporation of CNTs. The significant increase in the microhardness value obtained in this study can be attributed to better dispersion of the CNTs (as depicted in Figure 2) which promoted better mechanical properties even at lower sintering temperatures as compared to literature.

4. Conclusion

In this work, CNTs were uniformly dispersed in nickel aluminide matrix using low energy ball milling in a bid to preserve the structural integrity of the nanotubes. Sintering parameters were varied to determine their effect on the densification and subsequently hardness of the spark plasma sintered composites. The experimental results obtained clearly depict the importance of sintering parameters in the final mechanical properties of sintered composites. Both the density and the microhardness increased with increase in sintering temperature. Sample A sintered at higher temperature showed higher microhardness value of 8.23 GPa. While Sample B sintered at lower sintering temperatures exhibited lower densification and subsequently lower microhardness value of 6.78 GPa. In spite of the improved microhardness observed in this study in comparison with literature, higher microhardness values are still expected with higher sintering temperatures as can be seen from this study. However further increase of the sintering temperature led to the melting out of the admixed powders during sintering. Hence further work is still on going to overcome this challenge and thus sinter at higher temperatures and to investigate the fracture toughness of the NiAl-CNT composites.

5. References

[1] Liang W, Jiang Y, He Y, Xu N, Zou J, Huang B and Liu CT 2011 The corrosion behaviour of porous Ni_3Al intermetallic materials in strong alkali solution. *Intermetallics* **11** 1759-65.

[2] Sulka GD and Jóźwik P 2011 Electrochemical behaviour of Ni3Al-based intermetallic alloys in NaOH. *Intermetallics* **7** 974-81.

[3] Zhu S, Bi Q, Yang J, Qiao Z, Ma J, Li F, Yin B and Liu W 2014 Tribological behavior of Ni3Al alloy at dry friction and under sea water environment. *Trib.Inter.* **75** 24-30.

[4] Liu E, Jia J, Bai Y, Wang W and Gao Y 2014 Study on preparation and mechanical property of nanocrystalline NiAl intermetallic. *Mat.& Des.* **53** 596-601.

[5] Noebe RD, Bowman RR and Nathal MV 1993 Physical and mechanical properties of the B2 compound NiAl. *Inter. Mat. Rev.* **4** 193-232.

[6] Povarova KB, Drozdov AA, Kazanskaya NK, Morozov AE and Antonova AV 2011 Physicochemical approaches to designing NiAl-based alloys for high-temperature operation *Russ. Metall. (Metally).* **3** 209-20.

[7] Geist D, Gammer C, Rentenberger C and Karnthaler HP 2015 Sessile dislocations by reactions in NiAl severely deformed at room temperature. *J. of all & comp.* **621**: 371-77.

[8] Sarmasti AB, Yazdanirad M, Nouri Khezrabad M and Karbalaie M 2011 Effect of alumina particle size and thermal condition of casting on microstructure and mechanical properties of stir cast Al–Al2O3 composites. *Mat. Sci & Tech.* **11**: 1653-66.

[9] Munir KS and Wen C 2016 Deterioration of the Strong sp2 Carbon Network in Carbon Nanotubes during the Mechanical Dispersion Processing—A Review. *Crit. Rev. in Sol. St & Mat. Sci.* **5** 347-66.

[10] Liu ZF, Zhang ZH, Lu JF, Korznikov AV, Korznikova E and Wang FC 2014 Effect of sintering temperature on microstructures and mechanical properties of spark plasma sintered nanocrystalline aluminum *Mat & Des.* **64** 625-30.

[11] Cooke RW, Kraus NP and Bishop DP 2016 Spark plasma sintering of aluminium powders pre-alloyed with scandium additions *Mat. Sci. & Eng. A.* **657** 71-81.

[12] Chen B, Kondoh K, Imai H, Umeda J and Takahashi M 2016 Simultaneously enhancing strength and ductility of carbon nanotube/aluminium composites by improving bonding conditions. *Scri. Mat.* **113** 158-162.

[13] Kurita H, Estili M, Kwon H, Miyazaki T, Zhou W, Silvain JF and Kawasaki A 2015 Load-bearing contribution of multi-walled carbon nanotubes on tensile response of aluminum. *Comp. Pt A: App. Sci & Man.* **68** 133-39.

[14] Ameri S, Sadeghian Z and Kazeminezhad I 2016 Effect of CNT addition approach on the microstructure and properties of NiAl-CNT nanocomposites produced by mechanical alloying and spark plasma sintering. *Intermetallics* **76** 41-8.

[15] Groven LJ and Puszynski JA 2012 Combustion synthesis and characterization of nickel aluminide–carbon nanotube composites. *Chem. Eng. J.* **183** 515-25.

Acknowledgement

Authors are grateful to National Research Foundation (NRF), South Africa and Global Excellence and Stature (GES), University of Johannesburg for financing the research that produced this work.

IOP Conf. Series: Materials Science and Engineering **628** (2019) 012008 doi:10.1088/1757-899X/628/1/012008

Tensile properties of the reclaimed steel rail flash butt joint

Anuar Pawan[1,2], Zuraidah Salleh[3*], Akid H Zulkefli[3], Ya'kub Md Taib[3], Anizahyati Alisibramulisi[2,4] and Nik Roslin Masdek[3]

[1] Public Works Department, Jalan Sultan Salahuddin, 50582 Kuala Lumpur, Malaysia
[2] Faculty of Civil Engineering, Universiti Teknologi MARA, 40450 Shah Alam, Malaysia
[3] Faculty of Mechanical Engineering, Universiti Teknologi MARA, 40450 Shah Alam, Malaysia
[4] Institute for Infrastructure Engineering and Sustainable Management (IIESM), Universiti Teknologi MARA, 40450 Shah Alam, Malaysia

corresponding author's e-mail: a_kzue@yahoo.com*

Abstract. The development of the public transportation vehicles such as railway train had an incredible impact on the expansion of human mobility. In constructing the railways, quality assurance must be considered as it is very important to the project. The Malaysian Public Works Department (PWD) has undertaken the East Coast Line Track Rehabilitation Project between Gemas to Mentakab, which is estimated to cost RM220mil, these 123 km-long railway rehabilitation projects would use the reclaimed rails rather than the new rails. The reclaimed rail was manufactured according to the European standard EN 13674-1 and previously installed on site were examined and tested before could safely use in this track rehabilitation project. The reclaimed rails are analysed using ultrasonic inspection, hardness test, tensile test and micro examination to ascertain the safety values of mechanical properties such as tensile properties, hardness properties and microscopic morphology which regards to the composition of the welding joint and rail. Thus, to determine the reclaimed rails able to withstand forces applied on it without causing failure.

1. Introduction

The public transportation is a vital importance across many countries. The development of the public transportation vehicles such as railway train had an incredible impact on the expansion of human mobility. Railway line and its facilities are continuously improved to meet this transportation needs. Quality assurance during railway construction is very important for safe and timely completion of the project. With the development of high-speed and heavy-load railway, more stringent requirements are put forward for the comprehensive properties of steel rails [1]. As the demand for traffic increases, the railway company tends to operate railway at higher speeds carrying heavier loads, thus reliable damage tolerance design and effective maintenance methods must be established [2]. Malaysian Public Works Department (PWD) has undertaken the East Coast Line Railway Track Rehabilitation Project between Gemas to Mentakab, which is estimated to cost RM220million and these 123 km-long railway rehabilitation projects would use the reclaimed rails rather than the new rails [3]. The UIC54 reclaimed rails were operational track material recovered from site. The rail is T-type flat bottom with 54.77 kg mass per meter, was manufactured and previously installed using Flash Butt Welding (FBW) according to the European standard EN13674-1 [4]. Using reclaimed rails with existing welded joint requires understanding of weld strength performance as rail joint is the weakest link in the track. Welded joint has variation in material hardness and during operation, the differences in rail head level can cause distortion to wheel passage impacting additional forces onto track material [5].

One of the preferred jointing technique in railway construction is using Flash Butt Welding. FBW provides high productivity, cost effective and producing high quality welding [6]. A flash butt weld is performed either in production site or on-site using track mounted machinery equipped with high capacity transformers and hydraulic system. A typical Flash Butt Welding process is described in Figure 1 as follows:

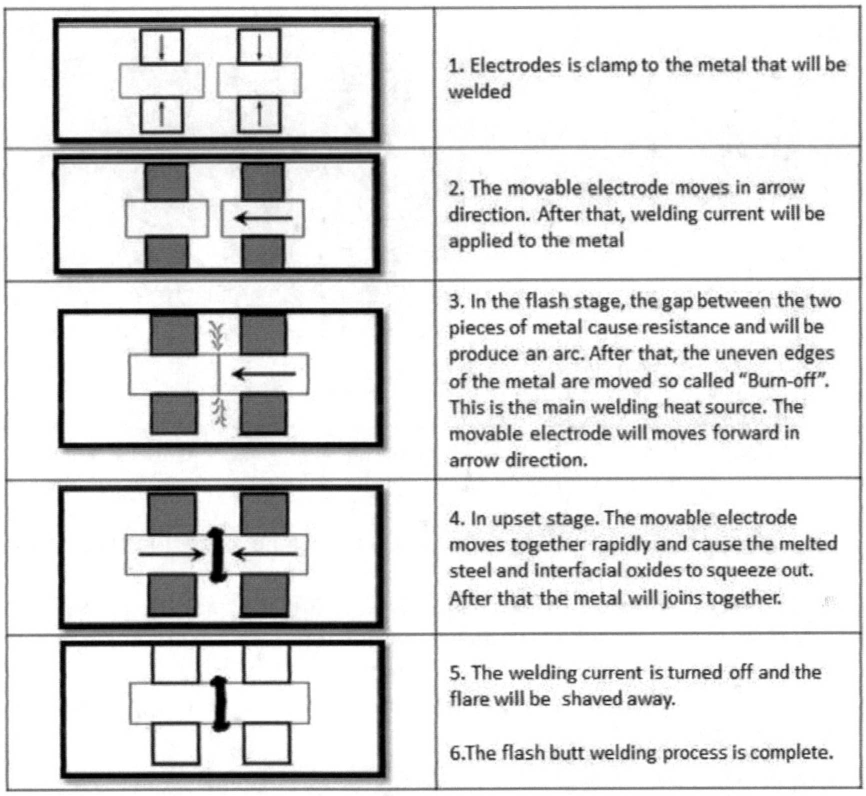

Figure 1. Typical FBW welding cycles from initial burn-off, preheating, flashing and retraction [6]

FBW process number 1 to 3 in Figure 1 shows the welding process begins with fixed end of one rail is pressed with another rail's end in the longitudinal direction. Then a clamping force is applied followed with high current passing through the clamps. The current heats up both rails causing both ends to melt and jointed as shown in Figure 1 (item 4)[6]. The rail feeding rate is programmed in between 0.1 to 2.5 mm/s and to ensure homogeneity of molten rail, a total length of 20 to 30 mm is consumed. Once desired length is obtained, the current is turned off and excess molten material will be shaved away using designated rail profile plate. Flash Butt Welding producing very high temperature in the fusion zone (FZ) and the heat affected zone (HAZ). These temperatures decrease rapidly as the distance from the fusion zone increases. This heating followed by cooling to ambient temperature produces stresses in the rail referred as residual stresses [7]. During railway operation, rails are subjected to temperature changes, cyclic loading, fatigue stress and inherent residual stresses might weaken the joints, thus pertinent investigation is required to ensure welded joints are in the good condition.

Thus, this study will determine the tensile properties, hardness properties and microscopic morphology which regards to the existing weld quality against base material experiencing the same operational

IOP Conf. Series: Materials Science and Engineering **628** (2019) 012008 doi:10.1088/1757-899X/628/1/012008

condition. Testing of rail sample is always following the relevant standards, regulations and legislations. The testing of rail welds is directly linked to the quality assurance of the railway.

2. Methodology

In this study, all specimens for tensile test, hardness test and metallographic analysis are taken from longitudinal section across the rail head as shown in Figure 2. For welded rail, the sample length is 260mm and covers several regions in longitudinal direction of the rail sample; base material (BM), heat affected zone (HAZ), fusion zone (FZ) and central line of weld (CL).

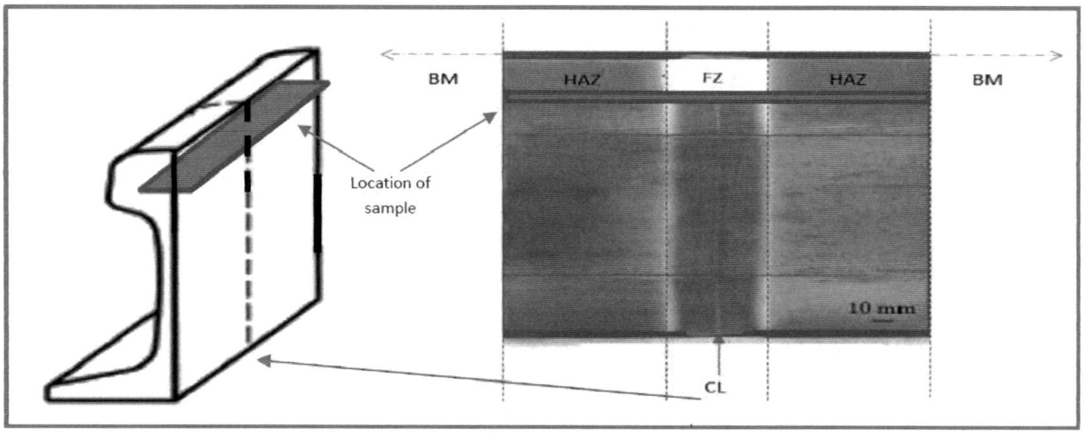

Figure 2. Location of samples are taken from UIC54 welded rail.

2.1. Sample preparations.

All specimens are inspected using ultrasonic inspection machine to ensure samples are defect-free from any cracks, flaws or internal material defects.

The sample preparation processes involve band-saw cutting, face milling and precision surface grinding. These processes are selected to eliminate additional heat into material, to ensure repeatability process of producing precise dimensioning and consistent surface finishes. Final test piece shapes are carried out using water-jet cutting process for high precision cutting dimensioning and low temperature abrasive cutting process.

2.2. Tensile test.

The standard used in preparation of tensile specimen and the testing procedure is adhering to ASTM E8/E 8M for consistent and reliable results [8]. Rectangular tensile specimens from the base material (BM) and welded rail samples were machined to standard 50mm gauge length and 12.5mm width. Tensile tests were carried out at room temperature at nominal strain rate of 0.3mm/min. The Universal Testing Machine (UTM) model Instron 3382 with load up to 100kN was used for tensile test.

2.3. Hardness test.

For hardness test, the specimens are prepared and tested in accordance to EN 14587-1. For both base material and the welded rail sample, the hardness distribution was measured with impression on line traverse from the central line (CL) to point on each side of the weld point 20mm into the BM; unaffected parent rail with spacing of 2mm centres. Vickers hardness HV30 was used in determining the hardness distribution across the BM and welded rail sample. The hardness testing machine model Mitutoyo MVK-H1 were used for hardness test.

2.4. Microstructure analysis.

The microscopic morphology analysis was determined by conducting microscopic examination according to EN 14587-1. This requires additional sample preparations involve thermoplastic mounting, polishing and etching. The analysis will examine the existence of the microstructures consisting of different sizes of ferrite, α-Fe and pearlite, α-Fe_3C from BM and welded rail and were observed from the micrographic analysis with a set number of magnifications. Microscope (Image Analyzer) Leica Q550 MW were used for microscopic examination.

3. Material properties

The type of reclaimed rail used in this study is rail profile UIC54 grade R900A. Table 1 shows the chemical composition and tensile properties of UIC54 grade R900A.

Table 1. The Chemical Composition of UIC54 rail (wt%) [4].

C	Mn	Si	Cr	S	P
0.6-0.8	0.8-1.3	0.1-0.5	-	0.04 max	0.04 max

The R900A rail has chemical composition (wt%) of 0.6-0.80 C, 0.8-1.3 Mn, 0.1-0.5 Si. This T-type section rail (flat bottom rails) with a 54.77 kg mass per meter with density of 7850 kg/m3. For a standard track, UIC54 rail is commonly used for medium and heavy load traffic [9,10].

4. Results and discussions

4.1 Tensile properties.

Tensile test specimens (ASTM E8/E 8M; Gage length: 50 mm, Width: 12.5 mm) are sampled from the head of the rail. The tensile shear test was conducted by using Universal Testing Machine (UTM) model Instron 3382 at room temperature with a constant nominal strain rate of 0.3 mm/min. Testing repeated for 3 samples of BM and welded rail. Comparison was made by observing the necking and fracture location on the specimens. Specimen J is unwelded and unaffected parent rail, hence the regions of CL, FZ and HAZ does not exist. Specimen K, L and M are rails joint by flash-butt welding. In general, specimen K, L and M experienced similar tensile behaviour. By observing Figure 3, specimen J experienced necking at a region in between the gauge length. Fracture also occurred in the same region with a value of elongation. While, specimen K experienced necking at two location which is HAZ. For sample K, the necking does not exist in FZ region as shown in Figure 4.

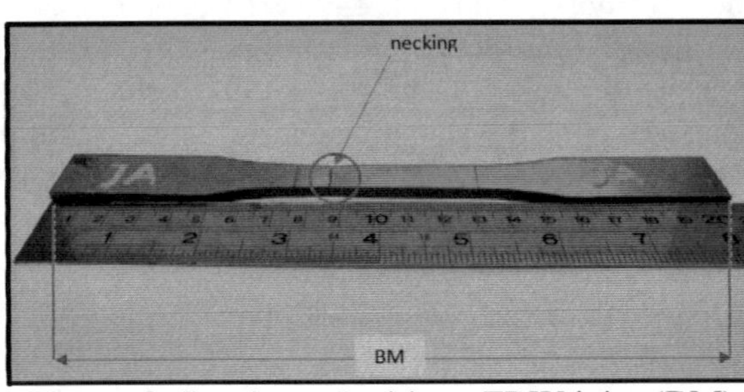

Figure 3. Specimen J without FBW joint (BM)

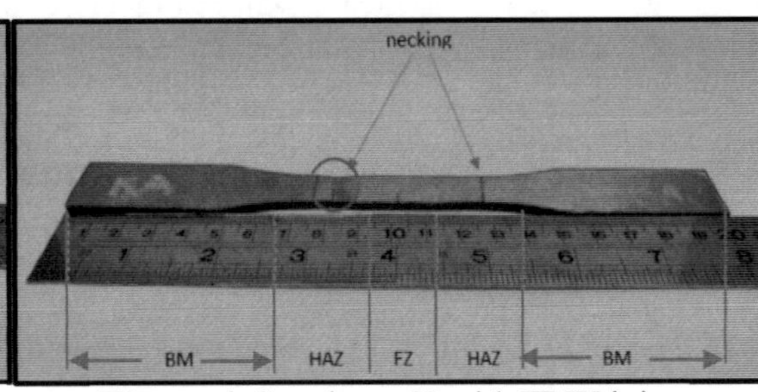

Figure 4. Specimen K with FBW joint

A brittle material usually ruptures when the maximum tensile normal stress exceeds the ultimate tensile stress. Since the samples elastic and plastic deformation as observed in the non-linear curve, the specimens are considered as ductile material as Stress-Strain Diagram in Figure 5.

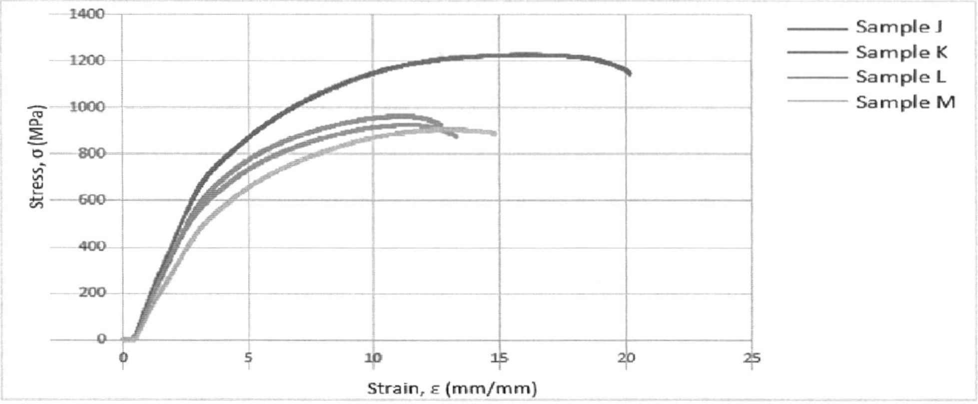

Figure 5. Stress-strain curve

Table 2 represents the tensile properties of both BM and welded rail yields the maximum load, tensile strain and UTS undergoes by welded rails have significant difference which is lower than BM. Referring to the non-linear deformation curve region in Figure 5, all samples experienced to strain hardening and limited period of necking.

Table 2. Tensile Properties

Tensile Properties	J	K	L	M
Maximum Load (kN)	46.04	33.72	34.79	35.34
Tensile Strain (mm/mm)	16.07	11.16	11.37	11.13
UTS (MPa)	1227.71	963.46	927.70	906.77
Young's Modulus (kN/mm)	18.48	16.31	17.11	16.12
Elongation (mm)	9.16	6.36	6.48	7.44
Ductility (%EL)	4.58	3.18	3.24	3.72

Only sample J experienced more necking and it was proven because sample J elongates with the value of 9.16 mm which is the greater value compared to other samples. Thus, base material is more ductile and lower brittleness rather than welded sample. It is also proven that the ductility of sample J is the greatest, refer Table 2. Hence, the application of FBW on the rail reduced the ductility and exhibits more brittle behaviour to the steel. Welded samples underwent FBW were decreased in the value of UTS, Young's Modulus and ductility. The percentage of decrement were calculated, and Table 3 shows all the tensile properties of the rail sample were decreased by average of 25% from its nature after the application of FBW.

Table 3. Mechanical Properties of rail steel after FBW

Tensile Properties	**Average value**		**Decrement**
	BM / Unwelded	**Welded**	**Percentage (%)**
UTS (MPa)	1227.71	932.64	24.03
Maximum Load (kN)	46.04	34.62	10.66
Young's Modulus (kN/mm)	18.48	16.51	26.20
Ductility (%EL)	4.58	3.38	24.80

The UTS of the welded sample dropped by 24% compared to BM. This shows the tensile strength of the rail steel were reduced and the rail steel becomes weaker. Decrement in ductility value explains that the welded rail steel exhibit little plastic deformation compared to BM as recorded on stress-strain

curve in Figure 5. The limited plastic deformation on the specimen exhibits brittle material and can be observed on non-linear curve of sample K, L and M in Figure 5 is smaller than sample J.
Generally, UIC54 has high uniform strain properties. Welded samples (K, L, M) have lower UTS value, less than 950 MPa compared to BM with UTS value of 1,227.7 MPa. The linear region shows the yield strength of the sample, BM have the highest yield strength compared to the welded sample. It can be stated that the tensile behaviour of UIC54 have been reduced from its nature if FBW is applied on the rail and HAZ is the weakest region on the track.

4.2 Hardness distribution.
The micro-hardness analysis was performed by using hardness testing machine model Mitutoyo MVK-H1 and comply to standard EN 14587-1. Hardness impressions were taken on a line 3-5mm below the running surface on the vertical longitudinal axis of the rail with spacing of 2mm between the points. An average of three readings at each location was recorded;

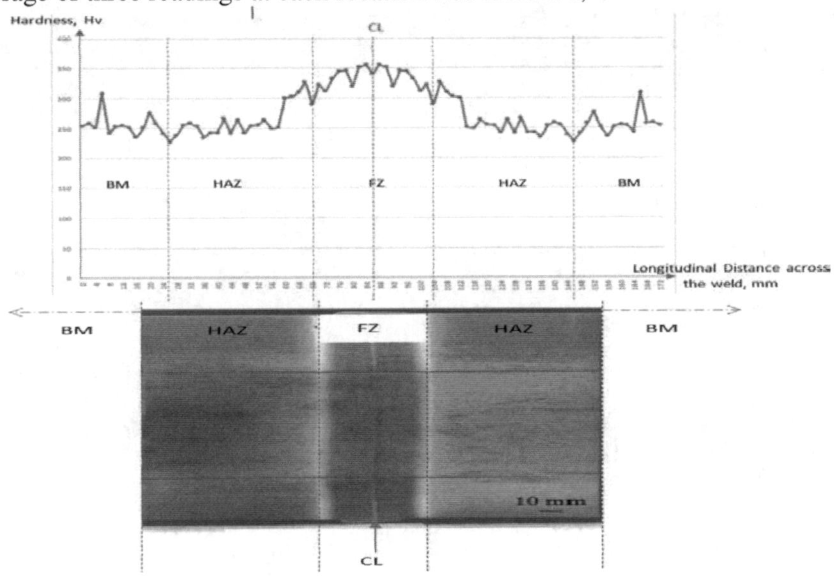

Figure 6. Hardness Distribution across the rail.

The hardness values are found distribute differently across the rail profile. The hardness vs. distance graph in Figure 6 is the typical trend of the welding hardness analysis [11,12]. Observing the hardness values in FZ, the hardness value is high in this region and decreasing when entering the HAZ. FZ region recorded the peak of the hardness which can relate to the tensile behaviour, where FZ resists the occurrence of the necking due to its high-hardness thus, fracture does not occur in FZ. As the hardness recorded traverse from CL, the reading is decreasing but, a significant increment shown from HAZ to BM thus, BM have high value of hardness compared to HAZ.
It can be stated that, after FBW process, the change in temperature and microstructure would alter the hardness behaviour of rail in FZ. However, hardness in HAZ is affected by the internal stresses during the cooling process. Rapid temperature drop causing reduction in hardness values similarly experience with accelerated cooling by means of blowing compressed air to the weld section. Further, at both sides of the HAZ, hardness is reduced; therefore, these zones are called the softened zone [13].

4.3 Microscopic Examination.
The microstructure analysis was performed by using Microscope (Image Analyzer) Leica Q550 MW and comply to standard EN 14587-1. The etching agent used have the following chemical composition per 10 litres; 1.875kg $CuCl_2$ – $2H_2O$, 5 litres HCl-1, 18ml - 35%, 4.2 litres distilled water. The same etching agent used for both BM and welded rail samples. Figure 7 to Figure 10 show the

IOP Conf. Series: Materials Science and Engineering **628** (2019) 012008 doi:10.1088/1757-899X/628/1/012008

microstructures consisting of different amount and size of ferrite, α-Fe and pearlite, α-Fe_3C. The pearlite microstructure exists as dark grains within each grain the layers are oriented in essentially same direction, which varies from one grain colony to another. The light constituents are the ferrite phase. Pearlite has properties intermediate between those of the soft, ductile ferrite and the hard, brittle cementite. Mechanically, pearlite is harder and stronger than ferrite microstructure [14].

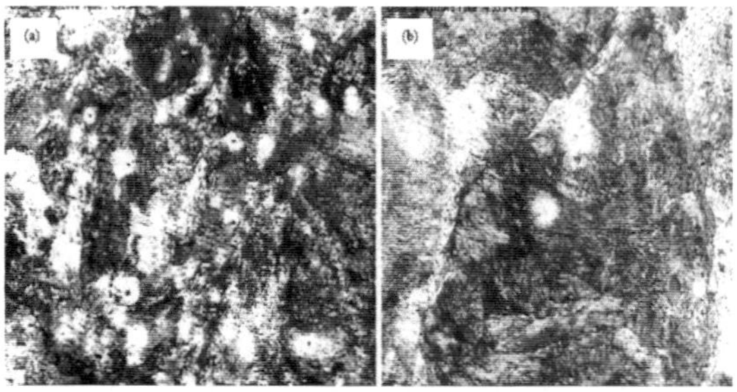

Figure 7. Micrograph of CL; (a)200x, (b)500x.

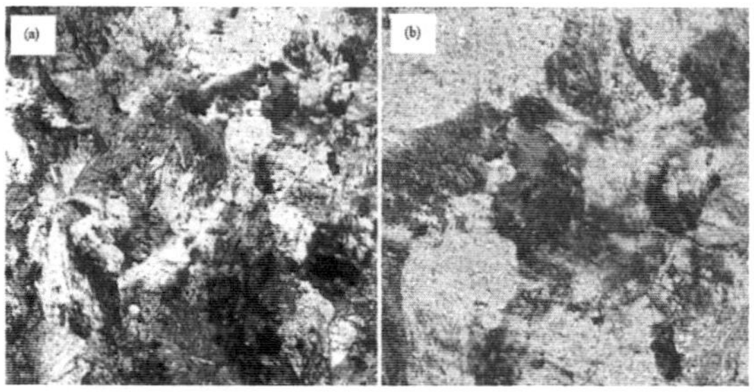

Figure 8. Micrograph of FZ; (a)200x, (b)500x

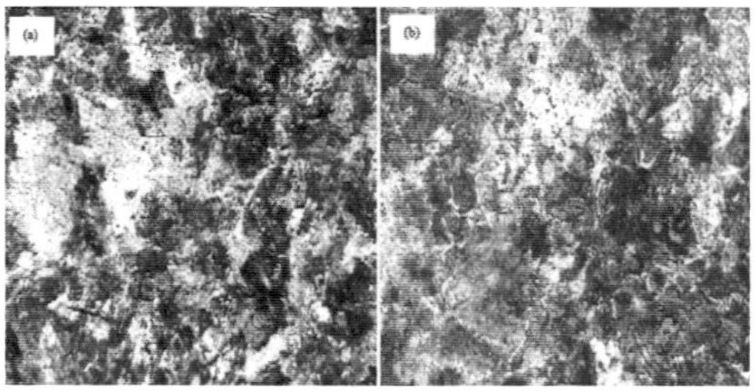

Figure 9. Micrograph of HAZ; (a)200x, (b)500x.

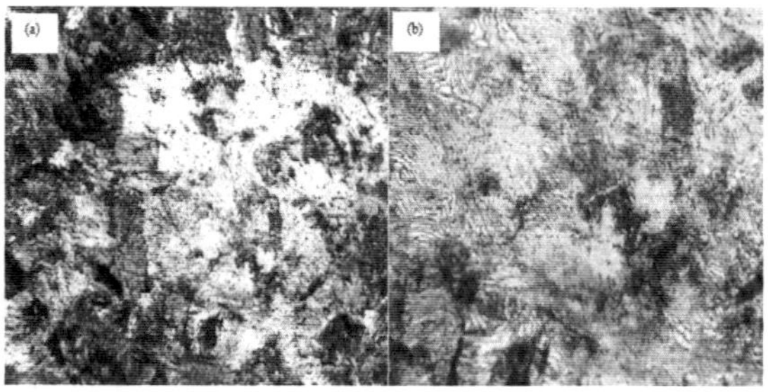

Figure 10. Micrograph of BM; (a)200x, (b)500x.

It is observed in Figure 7 that the micro-graph of CL has higher volume fraction of dark grains which is pearlite and limited light ferrite constituents. Thus, the CL region represents the behaviour of high hardness and strength properties. Figure 8 indicates the FZ with significant amount of ferrite and pearlite region as CL but coarser grain size. Coarse grain size associates with the low grain boundaries that acts as a barrier towards dislocation motion resulting the atoms in the grain will easily slides past one another [14]. Therefore, coarser grain size associates with hard, strong yet brittle. Fine grains of ferrite and pearlite can be seen in HAZ region as shown in Figure 9. Higher volume fraction of white constituent is observed represents a larger amount of ferrite microstructure. Mechanically, ferrite has properties of soft and ductile [14].

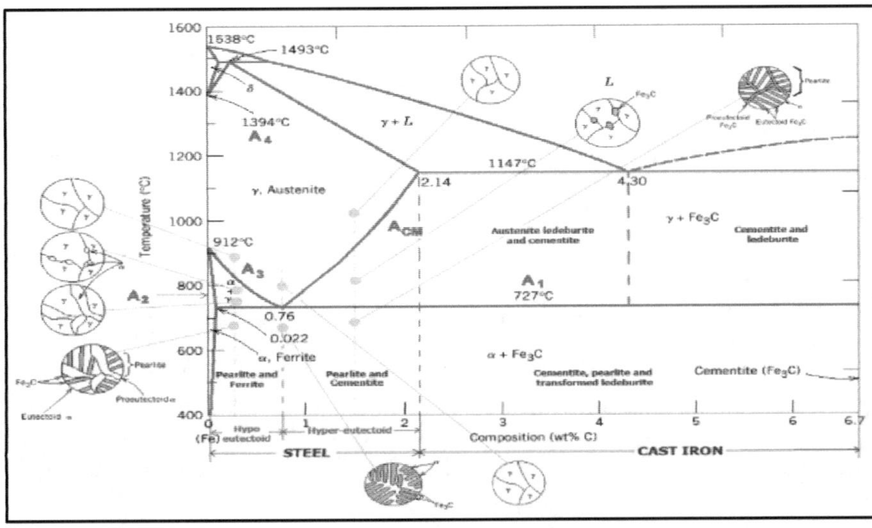

Figure 11. Iron-Iron Carbide (Fe-Fe_3C) Phase Diagram [14]

The R900A rail used in this study has chemical composition (wt%) of 0.6-0.80 carbon, C which is in the range of steel, refer to Figure 11. Composition of 0.6-0.80 wt% C is in between of the eutectoid-steel composition (0.76 wt% C) is consists only pearlite and ferrite. Therefore, high amount of ferrite in HAZ resulting the weak and soft behaviour but, high in ductility [14]. In the weld section, there are two distinctive regions. One is the region where the rail material is heated completely up to an austenite temperature region (dark part) by welding, and the other is the region where the rail material is heated to above the A1 point (approximately 727°C) and to a two-phase region (white part) that exists on both sides of the above-mentioned region. These two regions are collectively termed as the FZ and HAZ [13]. The occurrence of necking in HAZ values as discussed in tensile properties is due to the high ductility in HAZ region which allows the specimen to experience strain-hardening deformation in the non-linear deformation curve as shown in Figure 4.

5. Conclusion

The tensile properties of reclaimed rail joints were determined with the several testing conducted in purpose of relating the hardness properties and microstructure morphology. Following conclusions can be drawn from the investigations:

- The necking occurred in HAZ region when a uniaxial load is applied indicates the inherent of internal stresses when the rail was cooled to the ambient temperature right after the welding process. Under tensile stress, the material breaks within HAZ region and the necking occurred on both sides from the CL. The elongation, %EL experience by base material is higher compared to welded rail indicates the HAZ has intermediate ductility properties even it is soft and weak.
- Low hardness value exhibits in the HAZ region showing the properties of soft and weak yet ductile.
- Microstructure analysis shown that higher volume fraction of white constituent (larger amount of ferrite microstructure) exists in HAZ region, indicates the weak and soft properties of ferrite in HAZ region.
- BM has greater values of UTS, Young's modulus and ductility compared to HAZ due to microstructure alteration from welding process.

6. References

[1] D X Li 2005 Analysis of causes for rail breaking and preventative measures, *Railway Stand Des 3* pp 67–69.

IOP Conf. Series: Materials Science and Engineering **628** (2019) 012008 doi:10.1088/1757-899X/628/1/012008

[2] Y D Li, C B Liu, N Xu, X F Wu, W M Guo and J B Shi 2013 A failure study of the railway rail serviced for heavy cargo, *Case Studies in Engineering Failure Analysis 1*, pp 243–248.
[3] The Sun 2017 *Public Transport Initiative* Available at https://www.pressreader.com/ [Accessed: 9-September-2017].
[4] British Standard Institution 2011 *BS EN 13674-1:2011 Railway Applications - Track - Rail - Part 1: Vignole Railway Rails 46 kg/m and above* Available at: http://www.standardsuk.com/ [Accessed: 02-May-2018].
[5] S Chandra and M Agarwal 2007 *Railway Engineering,* (London: Oxford University Press), pp 294-311.
[6] K Saita, K Karimine and M Ueda 2013 Trends in Rail Welding Technologies and Our Future Approach, *Nippon Steel & Sumitomo Metal*, (Japan: UDC) pp 84–92
[7] G H Little and A G Kamtekar 1998 The effect of thermal properties and weld efficiency on transient temperatures during welding, *Computers & Structures*. **68**:1–3, pp 157–165.
[8] American Society for Testing and Materials 2016 *ASTM E8 / E8M-16a Standard Test Methods for Tension Testing of Metallic Materials* Available at https://www.astm.org/ [Accessed: 17-Jun-2017)
[9] ArcelorMittal Europe: Long-Products - Rails & special section, 54E1 (UIC54) Rail" [Online] Available: http://rails.arcelormittal.com/ [Accessed: 02-May-2018].
[10] British Standard Institution 2009 BS EN14587-2 *Railway application - Track- Flash butt welding of rails Railway applications. Infrastructure. Flash butt welding of new rails. R220, R260, R260Mn, R320Cr, R350HT, R350LHT, R370CrHT and R400HT grade rails in a fixed plant* Available at: http://www.standardsuk.com/ [Accessed: 02-May-2018].
[11] F A Ghazali, Z Salleh, K M Hyie, Y M Taib and N M Nik Rozlin 2017 Improvement of Mechanical Properties and Fatigue Failure of Spot-Welded Joint through Pneumatic Impact Treatment (PIT), *Pertanika J. Science. & Technology*. 25 (S), pp 105 - 114
[12] Z Salleh, Y H P Manurung, Y M Taib 2018 Improvement of Mechanical Properties in Treated Spot Welded Joint, *Journal of Mechanical Engineering*, **5**(6), (2018) pp 239-247
[13] K Saita, K Karimine and M Ueda 2013 Trends in Rail Welding Technologies and Our Future Approach, *Nippon Steel & Sumitomo Metal, Japan UDC*, pp. 84–92
[14] W D Callister and D G Rethwisch 2014 *Material Science and Engineering*, 9th edition, ISBN: 978-1-118-31922-2. (New York: Wiley–Interscience)

Acknowledgements
The authors gratefully appreciate the office of Public Works Department Malaysia, Ministry of Transport and Hikmat Asia Sdn Bhd, for their support in this railway material research project under Industrial Collaboration Program (ICP) for Gemas-Mentakab Track Rehabilitation Project managed by Technology Depository Agency Berhad (TDA Berhad). Technical and financial assistance provided by the Universiti Teknologi MARA (UiTM), in particular; Institute of Graduate Studies (IGS Support Fund), Faculty of Civil Engineering and Faculty of Mechanical Engineering are highly appreciated.

IOP Conf. Series: Materials Science and Engineering **628** (2019) 012009 doi:10.1088/1757-899X/628/1/012009

Kinetic study of the photocatalytic oxidation of ethylene over TiO2 thin films

Rodica-Elisabeta Stroe[1] and Lasse A. Rosendahl

Department of Energy Technology, Aalborg University, 9220 Aalborg, Denmark

[1]Corresponding author: res@et.aau.dk

Abstract. The photocatalytic oxidation process has been recognized as an environmental friendly, sustainable and low-cost technology for air treatment. A photocatalytic batch reactor has been employed to assess the kinetics of the oxidation of ethylene (C_2H_4) over immobilized titanium dioxide (TiO_2) and the effect of initial pollutant concentration on the kinetics. The rate model based on the simplified mono-molecular Langmuir-Hinshelwood (LH) kinetic model was found to represent the degradation satisfactorily, however with numerical constrains. These were overcome by employing two different approaches. On one hand the Langmuir-Hinshelwood model was solved by means of the Lambert W-function and on the other hand it was reduced to a pseudo-first order rate law, both resulting in a good fit to the experimental data, thus proving that the photodegradation of C_2H_4 over TiO_2 can be modelled by a first order rate law under low concentration conditions. Moreover, both models give a linear dependence between the reaction rate and the initial C_2H_4 concentration, making an estimation of reaction rate as a function of concentration possible, for a larger interval of C_2H_4 concentrations.

1. Introduction

In our present days resources are already scarce and it is expected for them to become even scarcer in the following years. This means that energy, water and even land cannot be misused, but must be used efficiently and sustainably. Yearly in Denmark alone food waste corresponds to a loss of approx. 1.5 billion euros, out of which fruits and vegetables spoilage, occurring during transportation, storage and retail, accounts for almost a quarter [1]. One of the most significant contributors to a high fruits, vegetables and flowers spoilage is the generation and build-up of ethylene (C_2H_4) in enclosed storage spaces. Ethylene is a hydrocarbon gas released by the aforementioned produce, which drives their ripening and growth, being responsible for changes in colour, texture and softening. A prolonged exposure or a high level of C_2H_4 causes these produce to ripe and degrade prematurely. By removing the C_2H_4 or minimizing its levels from the enclosed atmosphere of storage or transportation facilities the shelf life of perishable produce can be prolonged and thus, waste reduced. However, removing ethylene from the surrounding of C_2H_4-sensitive produce is one of the main challenges that the postharvest industry is facing. Even though this C_2H_4 generation problem stems from domains such as plant metabolism, biochemistry and postharvest biology, the solution can be found in chemical engineering, photocatalysis, material science, and adsorption mechanisms [2].

Photocatalytic oxidation (PCO) processes have gained popularity in the recent decades, due to the growing attention given to indoor air quality control. Such processes are effective at removing low traces of volatile organic compounds (VOCs), nitrogen oxides (NOx), and carbon monoxide (CO) at room temperature and pressure and may be more cost-effective than other technologies, such as

IOP Conf. Series: Materials Science and Engineering **628** (2019) 012009 doi:10.1088/1757-899X/628/1/012009

chemical scrubbers or activated carbon adsorbers, due to the fact that the catalyst is inexpensive and capable of mineralizing most VOCs [3], among which C_2H_4. However, this technique is still under the development stage and little research has been made using C_2H_4 as a model compound at concentrations in the ppm range, typical for storage facilities of fresh produce and greenhouses. Extrapolating the collected degradation data at concentrations much higher than those of interest may not be valid [4]. In this sense several studies have shown that the inlet pollutant concentration dictates the reaction rate, where increasing inlet concentration enhances the reaction rate and may even cause a shift in reaction order [5].

The comprehensive modelling of photocatalytic reactors is complex. This involves coupling the radiation field with the fluid dynamics and the appropriate reaction kinetics [6, 7, 8]. Most of the studies have successfully modelled the effect of radiation intensity and optical properties on the PCO process, nevertheless rigorous models which can predict pollutant degradation as a function of real operating conditions and UV intensity have received little attention, especially gas phase photocatalytic reactors.

Thus, this study aims at investigating the reaction kinetics of C_2H_4 over immobilized TiO_2 thin films at concentrations in the ppm range and deriving a model for predicting the effects of pollutant initial concentration on the reaction kinetics in a batch setup.

2. Experimental setup and method

Figure 1 depicts the schematic of the experimental setup used for the gas phase PCO of C_2H_4 and figure 2 shows the reactor chamber where the oxidation takes place.

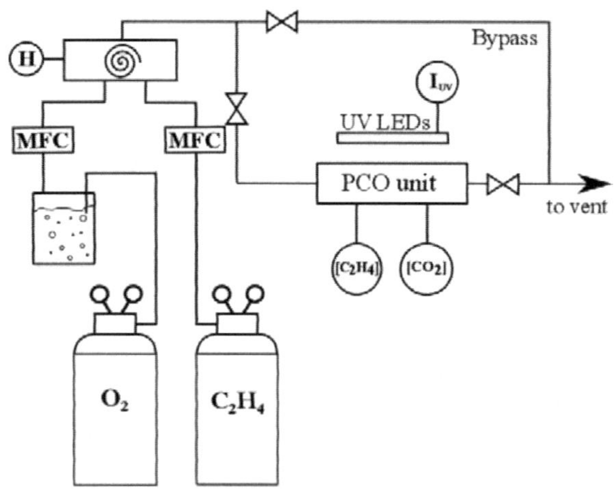

Figure 1. Schematic of the experimental batch setup.

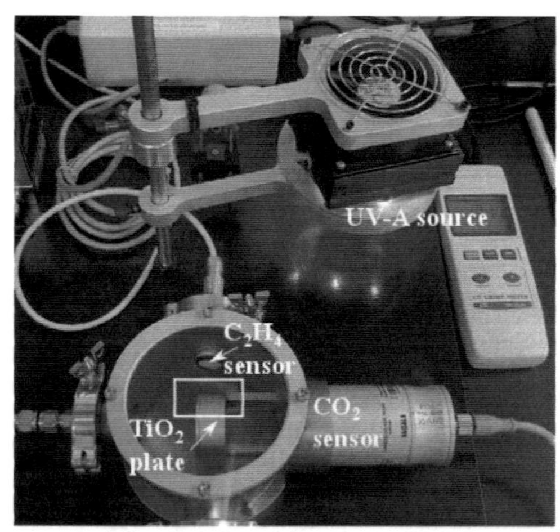

Figure 2. Reaction chamber – top view.

The cylindrical reactor has stainless steel walls, except for the top, which is made out of a 3 mm thick borosilicate glass to allow the irradiation of the catalyst. The reactor was irradiated with seven UV-A LEDs (LED ENGIN LZ-00UV00) each mounted on a 4.4mm x 4.4mm ceramic package with an integrated glass lens of Ø3.2 mm. The diodes have a radiant flux of 1.2 W UV-A and emit between wavelengths of 340 and 400 nm with a maximum irradiance peak at 365 nm. All seven LEDs are fixed on an Ebmpapst 3412 NHH heat sink. Their intensity can be easily varied by adjusting the voltage received. The UV radiation incident on the catalyst surface is measured via a UV light meter (LT Lutron YK-35UV). Since the catalyst surface cannot be shadowed the UV sensor and the chamber cannot be opened before experiments due to contamination risks, the UV probe was placed outside of the chamber, as seen in figure 2. The light intensity was measured by moving the UV-A source above the probe at the same distance to the catalyst. In addition to this, a second quartz glass is used to replicate the same medium the radiation needs to travel.

In addition to the inlet and outlet the reactor has two more openings where sensors are fitted to monitor the C_2H_4 and CO_2 concentrations. The C_2H_4-monitoring sensor is a nonselective PID AH-2 photo ionisation device which detects all VOCs with an ionisation potential lower than 10.6 eV, in a concentration range between 5 ppb and 50 ppm. In this setup using a nonselective sensor is not an issue, since it is assumed that the only detectable VOC present in the gas mixture is C_2H_4. The products of the oxidation are also monitored via a Vaisala GMP343 carbon dioxide probe.

The gas line consists of two streams – one containing the oxidizing agent, O_2, and one containing the pollutant, C_2H_4 (700ppm, N_2 balance), in the same carrier gas, N_2. The 79%N_2 + 21%O_2 gas mixture is bubbled through water to control the humidity for the reaction, after which the two gas lines are joined and mixed in a mixing unit, equipped with a Vaisala HUMICAP HMP110 humidity and temperature transmitter. The relative humidity (RH) is kept at 80% for all experiments, this being a typical value for storage units. The volumetric flow rates were adjusted by two mass flow controllers (MKS GE50A series) with a maximum rate of 500 cm^3/min. The desired concentration of C_2H_4 in the gas stream was adjusted by setting the volumetric flow rate at the needed value.

TiO_2 was deposited on aluminium substrate at a normal thickness of 1 μm by a pulsed DC magnetron sputtering using an industrial CemeCon CC800/9 SinOx coating unit.

Prior to the photocatalytic reactions, the TiO_2 thin plate was irradiated in pure air to oxidize any organic impurities on the catalyst. The concentration of pollutant was set by adjusting the flow rates accordingly. Then RH and C_2H_4 concentration in the reactor were allowed to reach a steady state. Once the steady state was reached the reaction chamber was closed off and the UV-A lamps turned on. The reaction was monitored until the concentration of C_2H_4 reached a constant value, dictated by the VOC sensor.

For performing the needed experiments the concentration of C_2H_4 is varied from 10 to 27 ppm, resulting in a flow rate range of 310 to 510 cm^3/min. The irradiance intensity and the relative humidity are kept constant at 60 W/m^2 and 80%, respectively.

3. Results and discussion

Langmuir-Hinshelwood (LH) kinetics is widely used to model the heterogeneous catalytic degradation during the photo-oxidation of organic pollutants in solution. Writing the LH kinetic rate law for a batch reactor yields the following expression:

$$r = -\frac{dC}{dt} = \frac{k_r K_{abs} C}{1 + K_{abs} C} \tag{1}$$

where r – reaction rate, [ppm s^{-1}], C – pollutant concentration, [ppm], k_r – reaction rate constant, [ppm s^{-1}] and K_{abs} – absorption constant of molecule on catalyst surface, [ppm]. In order for equation (1) to be further used in determining the reaction constants by fitting the model to experimental data sets, it needs to be integrated and linearized, obtaining equation (2).

$$\frac{1}{k_r K_{abs}} \ln\left(\frac{C_0}{C}\right) + \frac{1}{k_r}(C_0 - C) = -t \tag{2}$$

where C_0 – initial concentration of C_2H_4, [ppm].

This equation form reveals that C cannot be directly evaluated, due to the fact that it cannot be cleared from the first term, thus an explicit solution cannot be found. The challenge lies in the fact that the term C is both inside and outside of the logarithm. Hence, the Lambert W-function is employed to be able to isolate the concentration C from equation (2), although this increases the computational time significantly.

This numerical drawback can also be circumvented by either solving equation (1) directly using the relation between the reaction rate r and the concentration C or by transforming equation (1) into a pseudo-first order rate law. Considering the fact that the contaminated flows in most photocatalytic processes have a low molar concentration of pollutant, it can be assumed that $K_{abs}C \ll 1$. Hence, equation (1) can be rewritten into a fundamental first order equation, resulting in

$$-\frac{dC}{dt} = k_r K_{abs} C \quad (3)$$

The term $k_r K_{abs}$ is commonly defined as k_{app}, apparent first order kinetic constant; $[s^{-1}]$.
Figures 3, 4 and 5 show the experimental kinetic data and the predicted kinetics by first, direct LH and explicit LH rate laws, fitted by non-linear least square method. The predicted kinetic constant are listed in table 1.

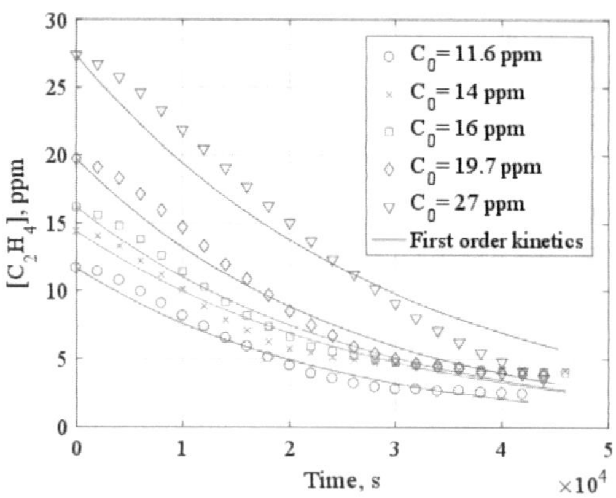

Figure 3. Experimental data vs. predicted C_2H_4 degradation by first order kinetics.

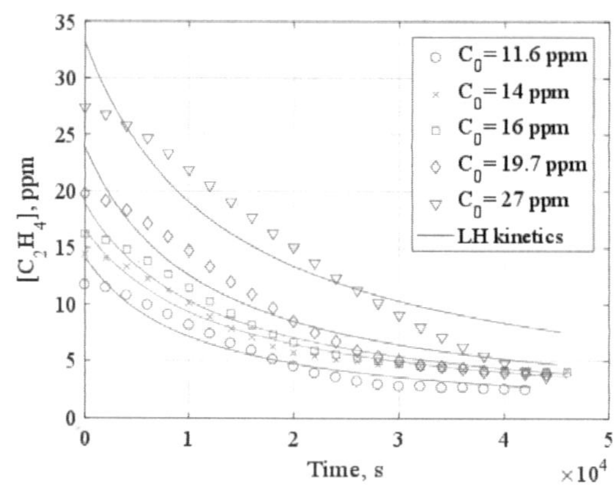

Figure 4. Experimental data vs. predicted C_2H_4 degradation by LH kinetics.

Table 1. Kinetic constants determined by non-linear least square method for the photocatalytic degradation of C_2H_4 over TiO_2, applying first order and LH kinetic model (W-function).

C_0 (ppm)	k_{app} (s^{-1})	k_r (ppm s^{-1})	K_{abs} (ppm^{-1})
11.6	4.304e-05	0.0014	0.0408
14	3.7003e-05	0.0839	4.4267e-04
16	3.8709e-05	0.0552	7.3064e-04
19.7	4.021e-05	0.0013	0.0477
27	3.434e-05	7.06e-04	0.3241

Figure 3 shows that the photodegradation of C_2H_4 over TiO_2 can be modelled by reducing the LH kinetic model to a pseudo-first order reaction rate, however there is deviation from experimental data that needs to be taken into account. Another drawback of using a first order kinetic model is the treatment of the two kinetic constants as one single term, thus a value for each constant cannot be determined, unless C_2H_4 adsorption isotherms are investigated to obtain a value for K_{abs}. Nevertheless, the model gives a linear dependence between reaction rate, r, and the initial concentration of reactant, C_0, as shown in figure 6, making an extrapolation to other concentrations possible, keeping in mind that an increase in C_0 might cause a shift in reaction order.
When solving equation (1) directly the initial concentration of C_2H_4 is not explicitly taken into account, as depicted in figure 4. Thus, the directly solved LH model fails to predict C_0, nevertheless it manages to estimate the end concentration, with generally less deviation than the aforementioned model. Although time consuming, as previously mentioned, using the Lambert W-function to isolate the C

IOP Conf. Series: Materials Science and Engineering **628** (2019) 012009 doi:10.1088/1757-899X/628/1/012009

term from the integrated form of the LH model, allows for the initial concentration to be considered, as readily seen in figure 5. Moreover, the fit of the model to the experimental data is superior to the previous two.
Figures 6a and 6b show the dependence of the reaction rate, r, to the initial concentration of C_2H_4, C_0. The reaction rate was calculated by using the kinetic constants experimentally determined and listed in table 1. Both data sets were fitted to a first order polynomial, the first order rate law having a better fit with an R^2 value of 0.94. The relatively poor fit of the LH rate law is caused by an outlier – the data point corresponding to C_0 = 27ppm predicts a decrease in rate, which is against the expected behaviour. Removing it increases the R^2 to 0.9, the relation between the two parameters becoming

$$y = 2.67e\text{-}05x + 1.488e\text{-}04 \tag{4}$$

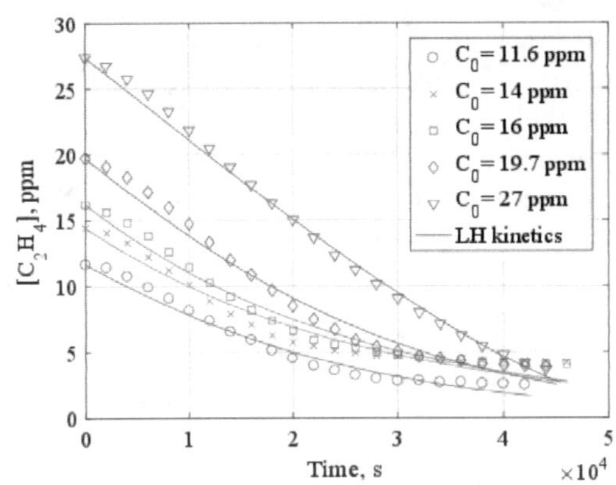

Figure 5. Experimental data vs. predicted C_2H_4 degradation by LH kinetics using Lambert W-function.

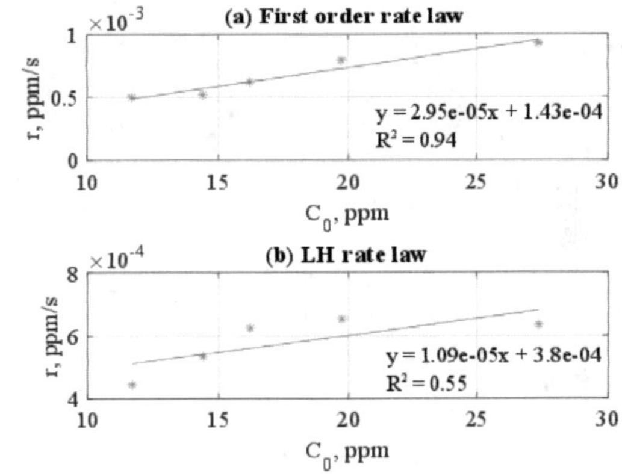

Figure 6. C_0 vs. r. (a) First order rate law. (b) LH rate law (Lambert W-function).

Often times not touched upon in literature are the numerical constrains that arise when employing the LH model and the correct way of circumventing these. It was readily shown in this study that the explicit evaluation of the C_2H_4 concentration is not possible with the LH model, unless a mathematical function, the Lambert W-function, is employed or the LH model is reduced to one of the fundamental rate models. Studies on PCO present in literature in which the LH model was reduced to a first order rate model lack the reasoning behind it and only invoke the assumption that for $K_{abs}C \ll 1$ the model reduction is reasonable. This is indeed true, as this study also has shown. The approach of assuming a first order rate law was proven to be effective and straightforward in estimating the degradation, however with minor deviations from the ideal first order equation, as also reported by [9], and experimental results (figure 3). The same approach was successfully used by [10] and [11], where the surface phenomena of TiO_2 photocatalysis and the photoelectrocatalytic degradation of C_2H_4 over TiO_2+activated carbon were studied, respectively. Thus, the reduction of the LH rate law to a first order type is in agreement with other studies made on photocatalytic reactions.
Although the use of the Lambert W-function slows down the computations, the estimation of the decrease in concentration is characterized by a lower squared norm of the residuals when fitted to experimental data, compared to the first order rate law (figure 5). This comes to show that by using the LH rate model correctly the degradation can be simulated with higher accuracy. A similar mathematical approach was employed by [12], although applied to a photocatalytic process using different model compound and catalyst. However, the same numerical constrains have been encountered and the Lambert W-function was employed to overcome these.

IOP Conf. Series: Materials Science and Engineering **628** (2019) 012009 doi:10.1088/1757-899X/628/1/012009

The linearity of the relationship between reaction rate and initial C_2H_4 concentration meet the original expectation, according to which the higher the C_2H_4 concentration the higher the availability of molecules at the active surface, resulting in an increase in reaction rate. In [13] a similar PCO system was analyzed in terms of kinetics – C_2H_4 over TiO_2, the authors deriving a linear relationship between initial concentration and reaction rate for concentrations ranging from 50 to 400ppm. As previously mentioned it is expected for the reaction order to shift towards a zero order as the concentration increases, and possibly also a change in dependence between concentration and reaction rate.

4. Conclusion

The kinetic analysis performed showed that the photocatalytic degradation of C_2H_4 over TiO_2 thin films can be well approximated by the most commonly used Langmuir-Hinshelwood kinetic model. Employing this model leads to an increase in computational time when performing a nonlinear fit to experimental data, due to the need of using the Lambert W-function to explicitly solve the model equation for the concentration *C*. It was showed that the LH kinetic model can be reduced to a pseudo-first order model, under the assumption that the pollutant concentration is low, such that $K_{abs}C \ll 1$. This allows to circumvent the use of the computationally heavy Lambert W-function. The disadvantage of reducing the LH model is the inability of estimating both k_r and K_{abs}, these being treated as one single term, k_{app}.

The determined reaction kinetics give a linear relationship between the initial concentration of C_2H_4 and the reaction rate, based on which the latter can be extrapolated for other initial concentrations outside of the experimental range.

5. References

[1] Minter M and Færgeman T 2011 Det skjulte madspild CONCITO (Copenhagen, Denmark)
[2] Keller N, Ducamp M N, Robert D and Keller V 2013 Chem. Rev 113 5029
[3] Wang S, And H M and Tade M O 2007 J Envint 33 694
[4] Obee T N and Brown R T 1995 Environ. Sci. Technol. 29 1223
[5] Zhao J and Yang X 2003 J. Buildenv. 38 645
[6] Salvado-Estivill I, Hargreaves D M and Puma G L 2007 Environ. Sci. Technol. 41 2028
[7] Qi N, Zhang H, Jin B and Zhang K 2011Chem. Eng, J. 172 84
[8] Passalia C, Alfonso O M and Brandi R J 2012 J. Haz. Mat. 211-212 357
[9] Asenjo N G, Santamaria R, Blanco C, Granda M, Alvarez P and Menendez R 2012 Carbon 55 62
[10] Ye S-Y, Tian Q-M, S X-L and Luo S-C 2009 J. Photochem. 208 27
[11] Fujishima A, Zhang X and Tryk D A 2008 J Surf Rep 63 515
[12] Kumar K V, Porkodi K and Selvaganapathi A 2007 J Dye Pig 75 246
[13] Einaga H, Tokura J, Teraoka Y and Ito K 2015 J Cej 263 325

Acknowledgments

The authors thank the Innovation Fund Denmark for financial support and the Danish Technological Institute for providing the experimental setup.

IOP Conf. Series: Materials Science and Engineering **628** (2019) 012010 doi:10.1088/1757-899X/628/1/012010

Characterisation and wear behaviour of rolled carbon steel in Dot 4 brake fluid

Olawale O Ajibola[1,2 *], Ojo J Akinribide[1], Samuel O Akinwamide,[1] Peter A Olubambi[1]

[1]Center for Nanoengineering and Tribocorrosion, School of Mining, Metallurgy and Chemical Engineering, University of Johannesburg South Africa
[2]Materials and Metallurgical Engineering Department, Federal University of Oye-Ekiti, Nigeria

Corresponding author: olawale.ajibola@fuoye.edu.ng

Abstract. The wear behaviour of as-rolled carbon steel (ARCS) in DOT4 brake fluid is investigated and reported in this study. Sample was characterised using chemical analysis, hardness, impact and tensile strength tests. Arc-spectrometry and the Energy Dispersive X-Ray (EDS) facilities were used to ascertain the chemical compositions. The microstructures and wear track patterns of samples were evaluated using Scanning Electron Microscopy (SEM) and High-resolution metallurgical microscope. The wear of as-rolled carbon steel (ARCS) samples subjected to reciprocating motion of loads at room temperature and constant wear time were compared in air, DOT4 fluid and water. The frictional behaviour and wear track patterns of the specimens subjected to varied normal loads (3, 5 and 10N) under reciprocating sliding wear were employed to assess the wear damage of samples in the three different environments. The coefficient of friction (CoF) obtained tends to be in the order of $\mu_{DOT4\ brake\ fluid} < \mu_{water} < \mu_{air}$ comparatively for the three test environments.

1. Introduction

Apart from the aluminium alloys, cast iron and steels are wide recommended for the manufacture of automobile and automotive engine parts such as the engine master cylinders and pistons; master brake and sliding calliper and piston components [1]. To get adequate information on behaviour of materials selected for use as components of automobile and automotive engines; diverse and intensive material characterisation and testing may be applicable. These may include the microstructural analysis, thermal and mechano-chemical behaviour of the materials when subjected to diverse manner of stresses and chemical environments [2-4].

Some automobile and automotive engine parts are subjected to various challenges such as the moderately high range of temperature and friction as in the combustion chamber of the engine block, cylinder and pistons [5]. In braking systems of automobiles, constant rubbing of contact surfaces often results in mild or intensive wear. A study shows that strong adhesive amalgamation appears at the tip of the asperities when two surfaces come in contact [6]. Therefore, adhesive wear particle emanates from shearing the boundary interface caused by sliding [7]. The hydraulic braking system is most famous among the various automobile braking systems. Friction contacts are produced by application of braking force with and without transmission fluids [8]. More research reports are available in the recent times on the characterisation, properties, tribology and corrosion behaviour of aluminium alloys

applicable to hydraulic brake cylinder pistons [9-10]. However, there is a lack of papers in the literature on the behaviour of steels used in similar service conditions and environments. The present work is a fraction of the extensive research work on the metal alloys - hydraulic fluids (DOT3 and DOT4) interactions, of which reports are available in public domains [11-18].
Wrought iron is a tough malleable, semi fused alloy of iron with fibrous slag inclusion, containing less than 0.08% C; suitable for forging and rolling in contrast to cast iron. Wrought carbon steel, also called the plain carbon steel is malleable with small Mn and some other elements. Rolled steel profiles of different shapes are produced by passing the stock through a single or multiple pair of rolls that reduces the thickness uniformly. The properties of the rolled products (mechanical, wear, weldability, corrosion) are much dependent on the interdependent set parameters which can be any or combination of rolling speed, rolling time and cooling rate. In a study by Adebayo and Ajibola [19], the influence of independent variables on the mechanical properties of rolled carbon steel was reported, while other studies investigated the influence of carbon content [20], cooling rate [21] and grain refinement [22] on the mechanical properties and microstructure of dual phase steels [23]. Steel rolling can be achieved hot or cold by similar processes, though the end products have different qualities. Thus the wear behaviour of as-rolled carbon steel (ARCS) specimen subjected to reciprocating motion of varied loads were compared in air, DOT4 fluid and water. The frictional coefficient and wear track patterns of the specimens under varied normal loads were used to assess the wear damages of the steel samples in the three selected environments.

2. Experimental procedures

2.1 Materials

The carbon steel materials used for the experiment were obtained from the light section rolling mill (LSM) of the Ajaokuta Steel Complex Ltd (ASCL) in Ajaokuta, Nigeria. The material was received as rolled 12mm diameter rods made from the as-cast 100 mm x 100 mm square billets. Arc-Spectrometer (2000-3 Spetro-CJRO model) and the Energy Dispersive X-Ray Spectrometer (EDS) facilities were used to determine the chemical compositions and it is presented in Table 1.

2.2 Sample Preparation and metallographic examination

The steel specimens were cut to 12 mm diameter by 10 mm thickness using the automatic precision cutting machine (Brillant 220, ATM, Germany). The surface was prepared to obtain even and flat mirror-like surface finish. The sample surfaces were water washed for 3 seconds and thereafter washed in methylated spirit to eliminate water stain on the surfaces before drying in warm shield air. The specimens were swabbed in Nital etchant containing 2 % vol. of Nitric acid and 98 % vol. ethanol solution (ASTM, E407-70) [24]. The microstructures were examined following standard procedures at different magnifications.

2.3 Mechanical testing and characterisation

2.3.1 Tensile test. Instron Universal tensile tester of type 3369 model was used to carry out the tensile tests. The test specimens were machined to the machine's specification in accordance with the ASTM standard. Each specimen was gripped at the two ends and then pulled until fracture and the ultimate tensile strength was recorded on the panel.

2.3.2 Impact test. The impact energy was determined by using Tinius Olsen 84-3 Model, Impact tester [18]. The test piece of a square cross-section of 10 mm dimension and 50 mm long, notched at the midpoint was used. The specimen was placed in a vice as a beam fixed at the two ends. The pendulum hammer of 75 kg weight was raised, on releasing; the specimen was hit behind the V-notch. The energy absorbed was read from the dial scale mounted on the machine and computed as a difference between the initial height h_i and final height h_f as the measure of the energy absorbed.

2.3.3 Micro-hardness test. The InnovaTest-(Falcon 500 model) was used for the hardness test. Three-point micro-indentations were made on the surface of the sample at regular interval of 3 mm distance apart. The dwell time was 10 s on HV/0.2 Vicker's scale.

2.3.4 Tribological tests. These sliding wear tests were carried out in two different environments. In the case of wet sliding tests, 20 ml of the fluid (DOT4 or water) is fed into the tribo cup, while the as-rolled carbon steel samples were immersed into the fluid (DOT4 brake oil or water) at atmospheric temperature of 25 °C and 50 % humidity. The tribometer was run in the reciprocating mode; sequence count of 10, frequency of 10 [Hz] with an acquisition rate of 100.0 [Hz] and fixed at a stop condition of 3 m. Three normal loads (3N, 5N and 10N) were applied via 6 mm alumina ball used to rub the surface of the rolled steel samples. The wear experiments were prepared, performed, analysed and reported considering the ASTM G 115-04, G 40-02; G 99 and G 133, DIN 50324, and ASTM G 99 - 95a. The CoF profiles were plotted against the sliding time. For clarity and easy view of reciprocating analysis, the CoF-sliding time curve on the two-way extraction profile was size filtered to 50 percent. The wear behaviour is evaluated based on the CoF (μ) obtained and compared in the three test environments.

2.4 Microstructural evaluation
The microstructures and wear track patterns of samples were evaluated using Scanning Electron Microscopy (SEM) (Vega3 Tescan model) and high-resolution metallurgical microscope (Zeiss MTB 2011 model). The frictional behaviour and wear track patterns of the specimens subjected to varied normal load under reciprocating sliding wear was employed to assess the wear damages of samples in the three different environments.

3. Results and discussion
The chemical compositions, mechanical properties and the three-point micro-hardness tests results of the as-rolled steel samples are shown in Tables 1-3.

Table 1. Chemical compositions of as-rolled carbon steels samples.

Elements	**Cr**	**Co**	**P**	**Cu**	**Si**	**C**
Amount %	0.69	0.016	0.016	0.276	0.101	0.357
Elements	**Mn**	**S**	**Ni**	**W**	**Fe**	
Amount %	0.63	0.015	0.08	0.158	97.661	

Table 2. Mechanical properties of as-rolled carbon steel samples.

Properties	***Value***
Tensile Strength, TS $_{@maxL}$ (N/mm^2)	673.18
Tensile Strength, TS$_{@Break}$ (N/mm^2)	494.83
Yield Strength (N/mm^2)	486.31
Hardness (HV)	168.74
Impact Strength (N/m)	100

Table 3. Three-point micro-hardness tests of rolled carbon steel sample.

Sample: H1	**P 1**	**P 2**	**P 3**
HV/0.2	179.4	163.5	163.5
Analysis: Max 179.4, Min 163.34, Average 168.74, Std.dev 7.53			
HBS500	155.4	141.8	141.7
MPa	607.2	552.3	551.8
Indentation d_1 (mm)	0.0483	0.0484	0.0477
Indentation d_2 (mm)	0.0427	0.0468	0.0476

3.1 CoF-sliding time patterns for ARCS under 3, 5 and 10N reciprocating loads

Table 4 and Figure. 1 illustrate the wear parameters of ARCS samples under the action of 3, 5, 10N reciprocating load in air, DOT4 fluid and water. There are variations in the COFs obtained from the test with respect to the sliding test parameters such as the normal applied load, track radius, slid time or cycle and wear environment/lubrication medium.

The CoF (μ) obtained tends to be the order of μ_{DOT4} (0.112~0.146) < μ_{water} (0.304~0.834) < μ_{air} (0.423~0.934) comparatively in the three test environments. Maximum COF values of 1.214, 1.118 and 0.242 were obtained from the test with water, air and DOT4 fluid respectively.

The graphical representations to compare CoF variation with applied normal loads in air, water and DOT4 fluid are presented in Figure 1. The CoF and CoF maximum values obtained in the sliding wear contacts are compared with respect to the three applied normal loads (3, 5 and 10N) and sliding environments (air, DOT4 fluid and water).

Generally, it is observed from the results (Figure 1) that the as-rolled carbon steel sample has tolerable friction behaviour in the three tested environments. Nonetheless, water shows less lubricating characteristic than the oil.

For the dry air sliding, the CoF fluctuates with increased applied load; it increased with load increase from 3N to 5N; thereafter diminished to a lower value. Increase in frictional coefficient for dry air sliding can be attributed to absence of surface wetting films. Whereas dissimilar CoF behaviour was observed in the wear analysis carried out in DOT4 environment. Decrease in frictional coefficient can be ascribed to resistance to wear resulting from the load bearing capacity of the film formed on the sample surface during wear. Sliding in water environment showed a clear trend of CoF increase as the load increased from 3 to 10N. Increament in CoF value can be as a result of deeper penetration due to increase in applied load. Further observations show a rapid decrease in frictional coefficient in ARCS analysed at various loads in DOT 4 environment in comparison to analysis carried out in dry air and water environments. ARCS have highest CoF in water as compared with air and DOT4 at 3N. Dry sliding in air gave highest CoF among the water and DOT4. Similar result was observed in a study by Duan et al. [25].

Table 4. Wear parameters of ARCS under 3, 5, 10N reciprocating load in air, DOT4 and water.

Load (N)	***Media***	***Cycle***	***Slid time (s)***	***CoF***	***CoF max***
3	Air	7059	716	0.770	0.923
3	DOT4	8168	816	0.112	0.228
3	Water	7147	714	0.834	1.214
5	Air	4088	408	0.423	0.852
5	DOT4	8412	841	0.146	0.242

IOP Conf. Series: Materials Science and Engineering **628** (2019) 012010 doi:10.1088/1757-899X/628/1/012010

5	Water	9529	953	0.332	0.789
10	Air	5997	540	0.934	1.118
10	DOT4	14345	1434	0.131	0.227
10	Water	14294	1429	0.304	0.627

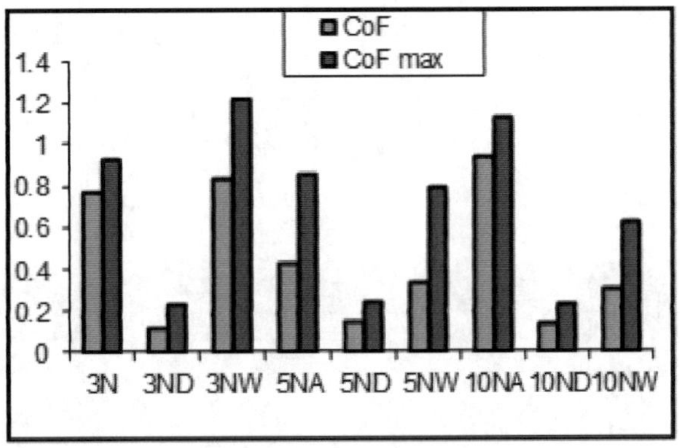

Figure 1. CoF variation with applied normal loads in air, water and DOT4 fluid.

3.2 Microstructures and wear patterns

Figure 2. shows the SEM and EDS of microstructures of ARCS test sample surface. Figure. 3. show the optical micrographs before and after dry sliding wear, while Figure. 4 are the SEM images of wear profile of ARCS impressed with alumina ball under the 10N load in air, water and DOT 4 sliding contact environments. The severities of the wear in the three environments are revealed in the images in Figures 3 and 4 as compared with Figure 2. Figure. 3c showed mild wear effect of wear in oil as compared with dry sliding (Figures. 3b and 4a) and wet sliding in water (Figure 4a). In Figure 3 the (a) microstructures and (b) dry sliding wear track of ARCS are presented. Strong adhesive union was formed at the tip of the asperities when two surfaces of ball and test sample are contacted and subsequently, adhesive wear particles were formed by shearing the interface caused by sliding in accordance with Akinwamide et al. [26]. In contrast, the strong abrasion is formed at the surface and region of contact between two load-bearing flat surfaces when ball meets the ARCS surface in water (Figure. 4b) and in oil (Figure. 4c) [7, 15-16]. Thus, abrasive wear particle is produced at the interface initiated by sliding.

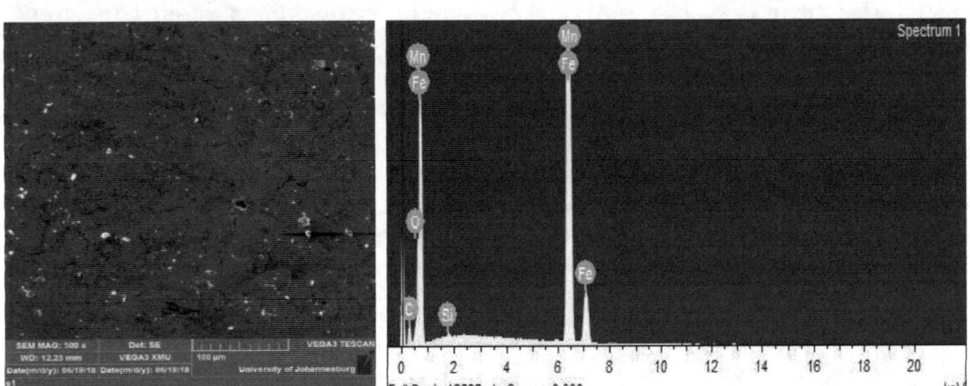

Figure 2. showing (a) SEM and (b) EDS of microstructures of ARCS.

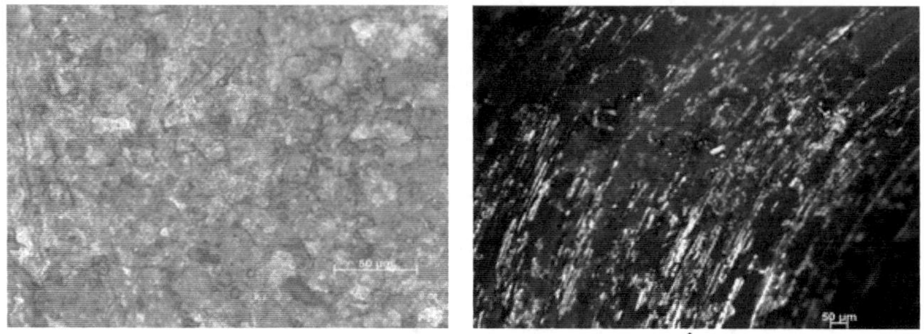

a b

Figure 3. showing optical micrograph of (a) microstructures and (b) dry sliding wear track of ARCS.

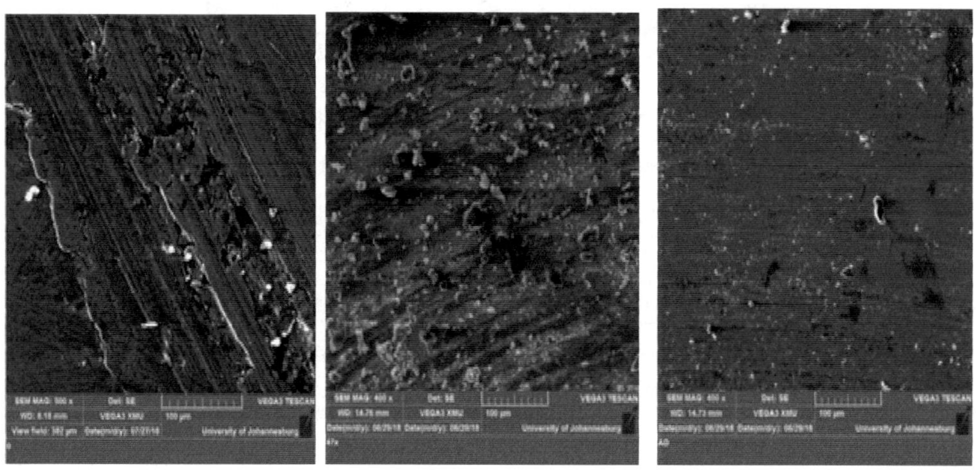

a b c

Figure 4. SEM showing wear tracks of ARCS under the (a) dry air (b) water and (c) DOT 4 sliding contacts.

4. Conclusion

The following conclusions are drawn from the results of the tests obtained; the CoF results show that the as-rolled carbon steel sample has the most tolerable sliding friction behaviour in the DOT4 fluid of all the three tested environments. Wear is most severe under dry sliding in air with strong adhesive union formed at the tip of the asperities when two surfaces are contacted. It is demonstrated that strong abrasion was experienced at region of contact between load-bearing ball with the ARCS surface sliding in water and DOT4 fluid. Notwithstanding, mild abrasive wear was experienced on the ARCS surface in oil.

5. References

[1] Pari, H., Raj, R., Pandiarajan, G. and Rasu, E., 2008. Study on the performance of electroless nickel coating on aluminium for cylinder liners. *Madras, India.*

[2] Grzesik, W., Zalisz, Z., Krol, S. and Nieslony, P., 2006. Investigations on friction and wear mechanisms of the PVD-TiAlN coated carbide in dry sliding against steels and cast iron. *Wear*, *261*(11-12), pp.1191-1200.

[3] Tetsui, T. and Miura, Y., 2002. Heat-resistant cast TiAl alloy for passenger vehicle turbochargers. *Mitsubishi Heavy Industries Technical Review*, *39*(1), p.200.

[4] Kragelsky, I.V., Dobychin, M.N. and Kombalov, V.S., 2013. *Friction and wear: calculation methods*. Elsevier.

[5] Singh, R.C., Lal, R., Ranganath, M.S. and Chaudhary, R., 2014. Failure of piston in IC engines: A review. *International Journal of Modern Engineering Research*, *4*(9), pp.1-10.

[6] Kinkaid, N.M., O'Reilly, O.M. and Papadopoulos, P., 2003. Automotive disc brake squeal. *Journal of sound and vibration*, *267*(1), pp.105-166.
[7] Louzguine-Luzgin, D.V., Ketov, S.V., Trifonov, A.S. and Churymov, A.Y., 2018. Surface structure and properties of metallic glasses. *Journal of Alloys and Compounds*, *742*, pp.512-517.
[8] Ajibola, O.O., Oloruntoba, D.T. and Adewuyi, B.O., 2014. Metallurgical study of cast aluminium alloy used in hydraulic brake calliper. *International Journal of Innovation and Scientific Research*, *8*(2), pp.324-333.
[9] Ajibola, O.O., Oloruntoba, D.T. and Adewuyi, B.O., 2015. Effects of moulding sand permeability and pouring temperatures on properties of cast 6061 aluminium alloy. *International Journal of Metals*, *2015*.
[10] Ajibola, O.O., Adebayo, A.O. and Oloruntoba, D.T., 2015. Corrosion of heat treated electroless-Ni plated mild carbon steels in dilute H2SO4'. *International Journal of Materials Science and Applications*, *4*(5), pp.333-342.
[11] Chate, G.R., Patel, G.M., Kulkarni, R.M., Vernekar, P., Deshpande, A.S. and Parappagoudar, M.B., 2018. Study of the effect of nano-silica particles on resin-bonded moulding sand properties and quality of casting. *Silicon*, *10*(5), pp.1921-1936.
[12] Green, R.L., Sauer, R.A. and Ford, G.P., Motors Liquidation Co, 1995. *High pressure brake hose with reinforcing layer of nonwater-based adhesive coated polyvinyl alcohol fibers*. U.S. Patent 5,445,191.
[13] Gorr, E. and Cameron, K., 2011. *Four-Stroke Motocross and Off-Road Performance Handbook*. Motorbooks International.
[14] Harvey, J.L., Milliken, L.T. and Forthofer, R.J., 1971. Trends in motor vehicle brake fluids and their standards. *SAE Transactions*, pp.1018-1036.
[15] Ajibola, O.O., 2016. Evaluation of Electroless-Nickel Plated Polypropylene under Thermal Cycling and Mechanical Tests. *Tribology in Industry*, *38*(3).
[16] Ajibola, O.O., Oloruntoba, D.T. and Adewuyi, B.O., 2015. Effects of moulding sand permeability and pouring temperatures on properties of cast 6061 aluminium alloy. *International Journal of Metals*, *2015*.
[17] Ajibola, O.O., Ige, O.O. and Olubambi, P.A., 2018. Wear and Corrosion of Wrought A6061 Aluminium Alloy in DOT3 Brake Fluid. *International Journal of Engineering & Technology*, *7*(2), pp.512-519.
[18] Ajibola, O.O., Adewuyi, B.O. and Oloruntoba, D.T., 2014. Wear behaviour of sand cast eutectic Al-Si alloy in hydraulic brake fluid. *International Journal of Innovation and Applied Studies*, *6*(3), p.420.
[19] A.O Adebayo. and Ajibola O. O. FUW Trend in Science and Technology Journal (FTSTJ) **2**,1 (2017)
[20] C. Adnan, D. Akin., S. Osman, U. Nazim, Verlag der Zeitschrift f¨ur Naturforschung, T¨ubingen (2010) http://znaturforsch.com. 0932–0784/10/0500–0468
[21] Nagie, J.M., 2014. The Effect of cooling rate on mechanical properties of carbon steel (ST 35). *Diyala Journal of Engineering Sciences*, *7*(1), pp.109-118.
[22] Muszka, K., Majta, J. and Bienias, Ł., 2006. EFFECT OF GRAIN REFINMENT ON MECHANICAL PROPERTIES OF MICROALLOYED STEELS. *Metallurgy and foundry engineering*, *32*(2), p.87.23.
[23] L. Valeria, N. Hernán, B. Lorussoa, G. Hernán, C. Svobodab, Procedia Mat. Sci. **8** (2015)
[24] Walker, P. and Tarn, W.H. eds., 1990. *CRC handbook of metal etchants*. CRC press.
[25] Lee, K., Dai, W., Naugle, D. and Liang, H., 2018. Effects of Microstructure of Quasicrystal on Their Mechanical and Tribological Performance. *Journal of Tribology*, *140*(5), p.051605.
[26] S.O. Akinwamide, S.M. Lemika, B.A. Obadele, O.J. Akinribide, B.T. Abe, P.A. Olubambi, Characterization and mechanical response of novel Al-(Mg–TiFe–SiC) metal matrix composites developed by stir casting technique, *Journal of Composite Materials*, (2019) 0021998319851198

Acknowledgement
The author would like to acknowledge the Management of Premier Wings Engineering Services, Ado Ekiti, Nigeria for providing the materials and some workshop services used for this work.

IOP Conf. Series: Materials Science and Engineering **628** (2019) 012011 doi:10.1088/1757-899X/628/1/012011

Fabrication, experimental investigation of jute fiber reinforced epoxy composites and hybrid composites

Tazbiul Mahmud Aranno [1], Sobahan Mia [2], Md. Shafayet Ullah Khan [1], M. A. Mehedi Hasan [1], Nyhan Shams Supto [1], Abu Shaid Sujon [3], Adib Bin Rashid [4], Md. Shafiul Ferdous [2]

[1] School of Mechanical Engineering, Bangladesh Army University of Science and Technology, Saidpur-5310, Bangladesh.
[2] Department of Mechanical Engineering, Khulna University of Engineering & Technology, Khulna-9203, Bangladesh.
[3] Department of Mechanical Engineering, Islamic University of Bangladesh, Gazipur-1705, Bangladesh.
[4] Bangladesh Military Academy, Bhatiary, Chattogram-4315, Bangladesh.

E-mail: munazeer_218@yahoo.com

Abstract. A special type of jute fiber woven mat was introduced by a hand loom. Epoxy jute fiber composites were fabricated using the woven fiber mat. A series of jute-hybrid epoxy composites based on the arrangement of the direction of jute fibers, and additional material were introduced to ensure the elevated mechanical properties of the composites. Improved mechanical properties were found in some cases of hybrid epoxy composites. It is discussed that the mechanical behavior is related to the fiber orientation, types of additional materials used in hybrid materials. The purpose of this study is to propose a simple technical method to prepare woven jute fiber mat by hand loom, and to produce epoxy jute fiber, hybrid composite materials to exhibit the elevated mechanical properties.

1. Introduction

The characteristics of the strength of the jute fiber epoxy composite, and hybrid composite were examined in this present study. Natural fiber is the one of the most popular biodegradable materials used for reinforcement. Therefore, the demand of fiber reinforced composites has been increased dramatically. The main reason behind this is high performance, great flexibility and handling advantages at favorable costs by shuffling of different matrix and fibers [1]. Presently, the scientists are playing their attention to the environmental impact of the existing synthetic fiber. They are trying to replace this synthetic fiber with the biodegradable, renewable and low cost natural fibers for the fabrication of composite material [2]. Composites are considered as a necessary material with high strength to weight ratio. Light weight and stiffness of composite materials play an important role in replacing the conventional materials [3]. However, study related to the development of these kinds of natural fiber reinforced composites to improve the mechanical properties is still inadequate. This study proposed a simple technical method to prepare woven jute mat by hand loom processes, and to fabricate composite material, hybrid material to get elevated mechanical properties.

2. Fabrication procedure

Clean and dry jute fibers are used to fabricate the epoxy jute fiber composites. Epoxy crystal resin and hardener are used as a ratio of 11:1 by weight as a matrix to fabricate these composite materials, and mixed homogenously for 5 to 10 minutes before starting the fabrication process. Among all the fabrication processes, Vacuum assisted Resin Transfer Molding (*VaRTM*) which impregnates dry-fabric performs with resin, and resin infusion technique is the most popular and widely used in industrial sectors, especially for carbon/carbon composite fabrication processes [4, 5]. But here hand layout with cold press method is proposed as a low-cost, reliable and rapid manufacturing process comparing with the other processes. Most of the researchers are using twill woven type mat which is formed by yarn using jute fiber [6-8]. Each yarn contains a bundle of fibers and the diameter of the yarn is about 0.6 mm to 1.0 mm. In those cases, the resin cannot penetrate inside the fibers of the yarn. Therefore, lack of interfacial adhesion between jute fiber and epoxy matrix occurs, i.e., the strength of the fabricated composites, reinforcement of the fiber and matrix become weak. Hence, to overcome from the drawback of yarn type woven composites, a new type of woven mat of jute fiber has been introduced to fabricate the composites using hand loom processes, where all the jute fibers used are free thus resin can penetrate each and every part of the woven mat. Unidirectional woven mat was produced by hand loom process. The physical appearance of the jute woven mat, fabricated jute composites using those mats, and the schematic diagram of the woven mat is shown in fig. 1. The thickness of each woven mat was about 0.8 mm. According to the arrangement of the fiber direction of the woven mat, two types of epoxy jute fiber composites were fabricated. Some hybrid composites were also fabricated using aluminum foil, glass fiber along with the jute fiber. Anggit et al. shows that the use of jute fiber for reinforcement is possible for wind turbine application [9]. The corrosive effect, flexural test, free vibration test, and some other parameters will be investigated in the near future to check the feasibility of the application of jute and hybrid composites in various industrial sectors.

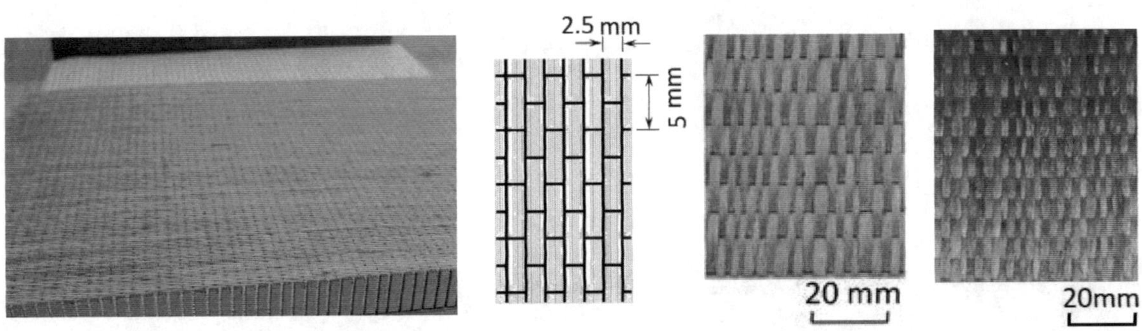

(a) (b) (c) (d)

Figure 1: Woven mat of jute fiber and composites: (a) Hand loom, (b) Schematic diagram of jute woven fiber mat, (c) Woven jute fiber mat fabricated by hand loom, (d) Epoxy jute composites after fabrication.

A steel frame, thin layer of polyethylene sheet and surface plate of about 220 kg weight was used to remove the excess resin and hardener mixture, air void between the fiber and resin, homogeneity of the material, and for the compactness of the fabricated composite materials. First a releasing agent and a polyethylene sheet were put on the mold to avoid the stickiness of epoxy resin to the surface of the frame. Then resin and hardener solution were applied to the polyethylene surface. Then a layer of woven jute fiber mat was laid down, followed by a quantity of resin, hardener solution onto it with the help of hand roller. The process was performed repeatedly until the required number of layers or required thickness of the composites were obtained. The surface plate is put on the top after completing the fabrication process, and kept for 24 hours at room temperature to get the perfect material. The schematic diagram of the jute composites fabrication by the manual hand layout process is shown in the fig. 2. In order to get the elevated mechanical properties of the composites, various types of jute hybrid epoxy composites were also fabricated. Aluminum foil of 41.5 GSM, glass fiber of 50.0 GSM was incorporated with the woven jute mat of 272.5 GSM to fabricate jute hybrid composites in various arrangements to get the improved mechanical properties. According to the arrangement of these materials, *A-F* types of composite materials were fabricated which are shown in table 1.

Table 1: Various types of jute composites and hybrid composites.

<table>
<tr><th>Types</th><th>Arrangement of materials</th><th>Types</th><th>Arrangement of materials</th></tr>
<tr><td>A type</td><td>|+|+|+|</td><td>D type</td><td>|+ Al + —— + Al +|+ Al + ——</td></tr>
<tr><td>B type</td><td>|+ —— +|+ ——</td><td>E type</td><td>|+ GF +|+ GF +|+ GF +|</td></tr>
<tr><td>C type</td><td>|+ Al +|+ Al +|+ Al +|</td><td>F type</td><td>|+ GF + —— + GF +|+ GF + ——</td></tr>
</table>

[a] Note: "Al": Aluminum foil; "GF": Glass fiber; vertical and horizontal lines indicate vertical and horizontal orientation of jute fiber in woven mat.

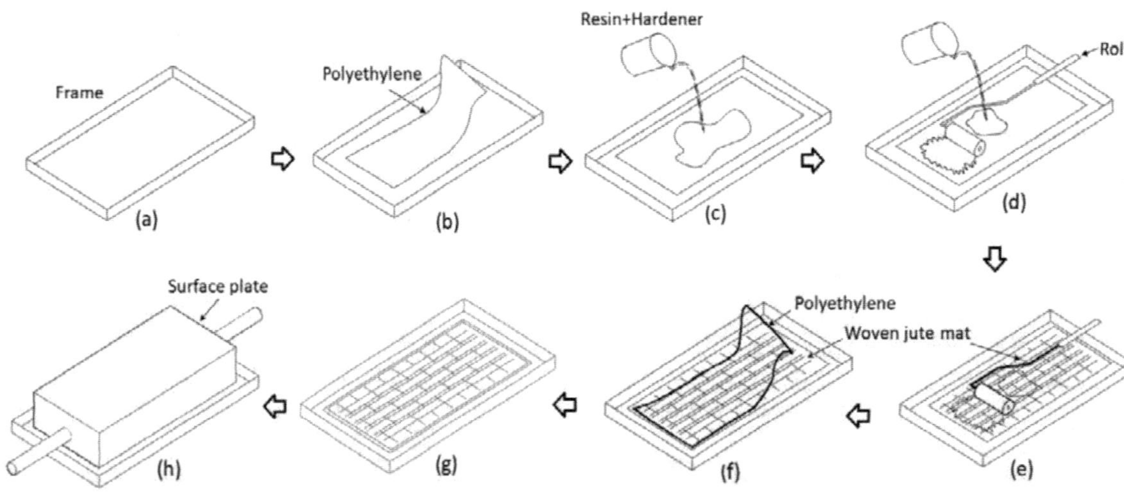

Figure 2: Schematic representation of the fabrication process of epoxy jute composites.

3. Experimental results and discussion

3.1 Impact strength

Charpy test was carried out to measure the impact strength of various types of fabricated composites. The specimen geometry for impact strength testing is shown in fig. 3 (b). Corresponding impact strength is shown in fig. 4. From the fig. 4, it was found that the maximum impact strength was found when unidirectional woven fiber mat was used to fabricate the epoxy jute composite materials (*A* type). When jute woven mat with alternating direction was used, the impact strength reduced considerably. From the experimental results, it was found that unidirectional fiber orientation of jute has higher impact strength rather than that of successive woven fiber mat with alternate direction used in composite materials, and addition of other foreign material to fabricate hybrid composite. As *A*, *C* and

E types of composites have all unidirectional jute fiber, the impact strength is higher than that of *B*, *D* and *F* types epoxy composite.

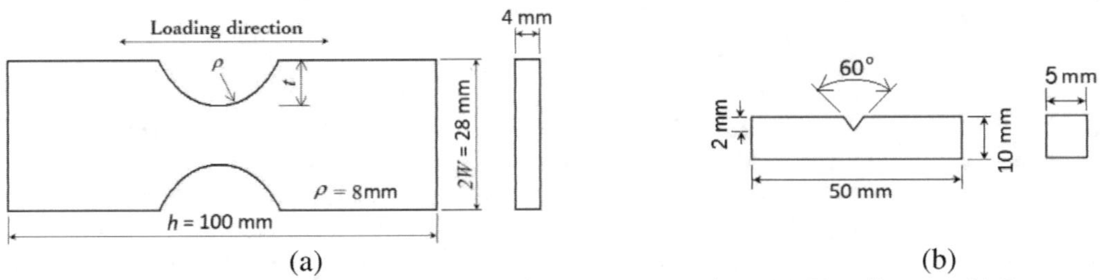

Figure 3: Specimen geometry of fabricated epoxy composites: (a) Tensile test, (b) Impact test.

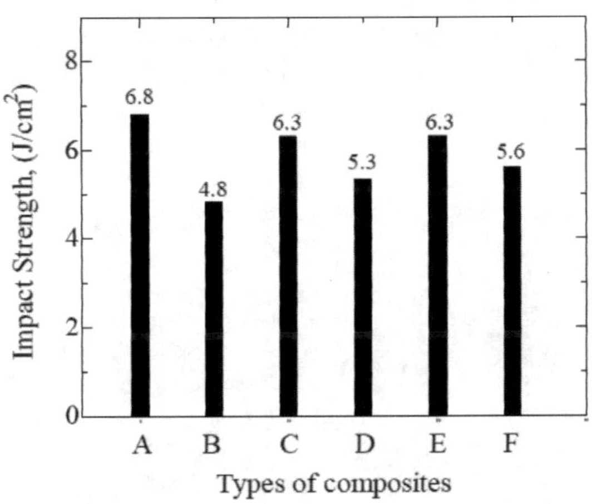

Figure 4: Result of Impact test (Charpy).

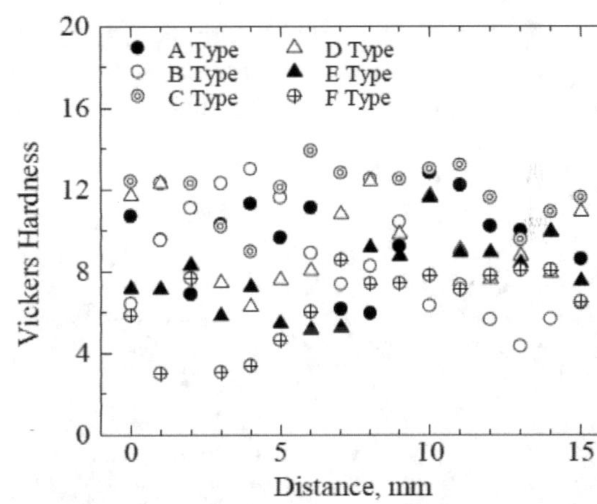

Figure 5: Result of Vickers hardness test.

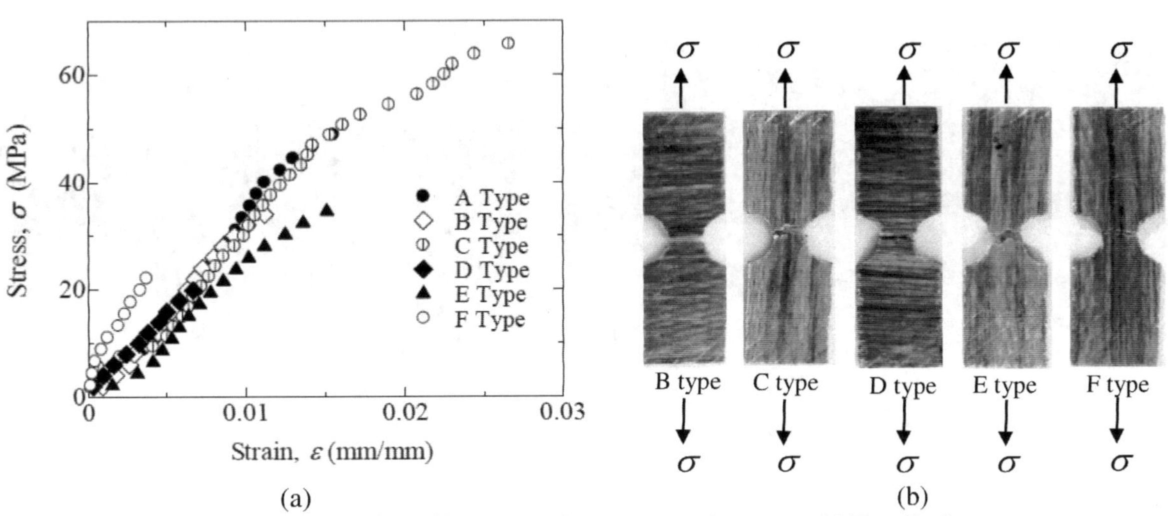

Figure 6: Result of tensile test: (a) Stress vs. strain curve, (b) Tensile fracture.

3.2 Vickers hardness

Figure 5 shows the Vickers hardness of epoxy jute fiber composites and hybrid composites materials. The hardness distribution was scattered. Maximum value of the average hardness (11.86) was observed in the case of *C* type material, and the minimum value of average hardness (6.39) was observed in the case of *F* type hybrid composite materials. As 5 layers of aluminum foil were incorporated in the *C* type specimen, Vickers hardness became higher than that of other types of specimen. The hardness distribution would also be related to the stress vs. strain curve.

3.3 Tensile strength

Figure 6 (a) shows the stress-strain curve of fabricated jute and hybrid composites. A series of tests were carried out to investigate the tensile strength. It was found that the tensile strength was higher when the direction of all jute fibers was in the loading direction (*A* type) comparing with the composites having woven jute fiber mat in alternating direction (*B* type). Lowest tensile strength was found in the case of *F* type hybrid composites where glass fiber was incorporated with jute fiber mats as shown in table 1. Also, the elevated tensile property was found when aluminum foil was incorporated with the jute fiber whose fiber orientation was in the loading direction (*C* type). From the experimental results, it was clearly found that the tensile strength of the *C* type specimen was improved due to the higher strength of aluminum foil and the unidirectional jute fiber woven mat used in the hybrid composites. Figure 6 (b) shows the tensile fracture of fabricated jute fiber epoxy composites and hybrid composites.

3.4 Microstructure by SEM

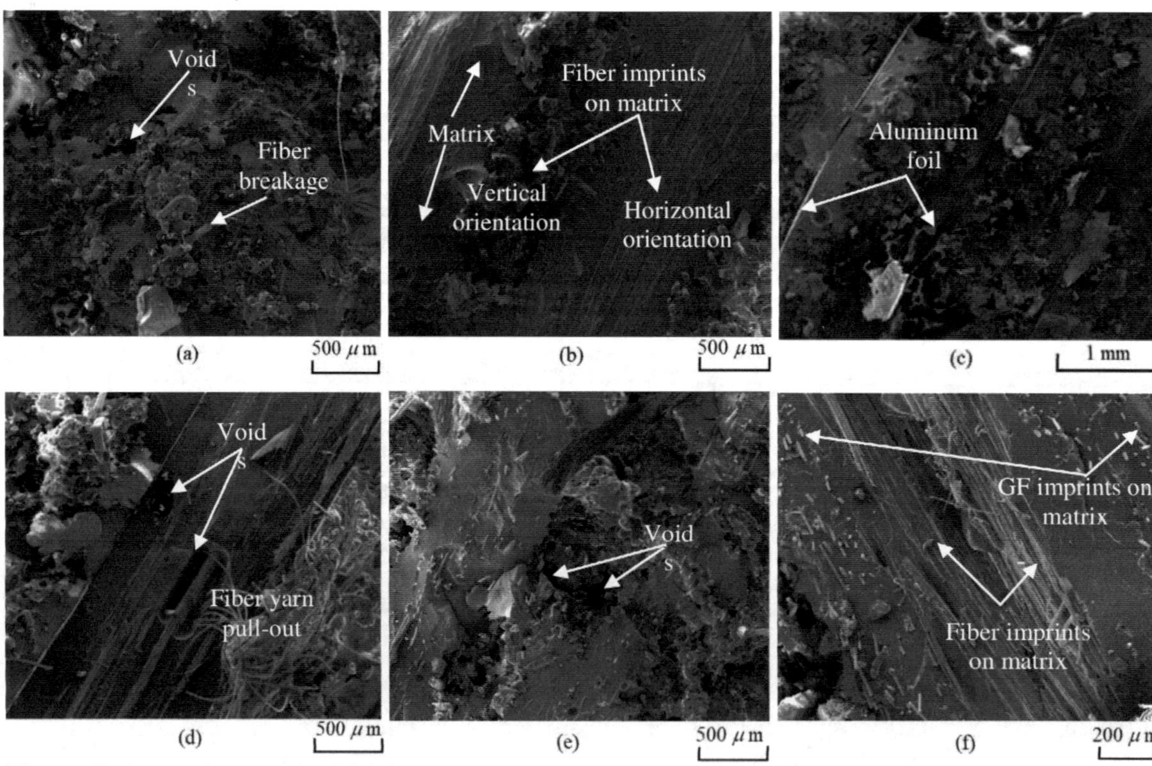

Figure 7: *SEM* micrograph of fabricated reinforced epoxy jute composites using jute fiber woven mat by *SEM*: (a) *A* type, (b) *B* type, (c) *C* type, (d) *D* type, (e) *E* type, (f) *F* type

Figure 7 shows the microstructure of various types of jute fiber reinforced epoxy composites and hybrid composites. The microstructure was taken by scanning electron microscope (*SEM*). In figure, the interfacial bonding between the jute fiber and matrix, distribution is clearly visible. It was also observed that the fiber imprints on the matrix when jute woven fiber mat by hand loom was used to fabricate the composites. Also, some air voids were found. These voids could be reduced or eliminated by improving the fabrication processes. The lack of adhesion between the pulled out fibers and the resin-hardener solution is visible in some cases (Fig. 7 (d)).

4. Conclusion

The mechanical properties of woven jute fiber composites and hybrid composites were investigated. The main results obtained are as follows:

(1) Higher mechanical properties of jute composites were obtained using woven jute fiber mats by hand loom processes, when the orientation of all the jute fibers was in the loading direction. Elevated mechanical properties were found for hybrid material where aluminum foil was incorporated with the jute fiber.

(2) The jute fiber perpendicular to the loading direction does not possess higher mechanical properties than that of jute fiber having loading direction. Therefore, the ratio of fibers in the loading direction is expected to be the one of the major parameters to improve mechanical properties.

5. References

[1] Minglel M, Guiling W, Baoyu L, Hongjuang M, *Engineering Review.* **32** (2012) 165-171.
[2] Rosa I. M. D, Santulli C, Sarasini F, *Mat. and. Design.* **31** (2010) 2397–2405.
[3] Saheb D. N, Jog J. P, *Adv. in Polym. Tech.* **18** (1999) 351–363.
[4] Ferdous M. S, Rahman S. M. M, Makabe C, *ICDMM 2018.* 11-13 August (2018) Okinawa, Japan.
[5] Makabe C, Fujikawa M, Ferdous m.S, Setyabudi S. A, *J. of High Pr. Ins. of Japan.* **51** (2013) 7-15.
[6] Bledzki A. K, Gassan J, *Prog. in Polym. Sci.* **24** (1999) 221–274.
[7] Eichhorn S. J, Baillie C, Zafeiropoulos A. N, Mwaikambo L. Y, Ansell M.P, Dufresne A, Entwistle K. M, Herrera-Franco P.J, Escamilla G. C, Groom L, Hughes M, Hill C, Rials T. G, Wild P. M, *J. of Mat. Sci.* **36** (2001) 2107–2131.
[8] Razera I. A. T and Frollini E, *J. of A. Polym. Sci.* **91** (2004) 1077-1085.
[9] Murdani A, Hadi S, Amrullah U. S, *Key Engg. Mat.* **748** (2017) 62-68.

Acknowledgment

Finally, the authors are thankful to Strength of Materials Laboratory of the University of the Ryukyus, Japan for their support in this research.

IOP Conf. Series: Materials Science and Engineering **628** (2019) 012012 doi:10.1088/1757-899X/628/1/012012

Wear behaviour of cast aluminium silicon (Al-Si) alloy in Dot 4 brake fluid

Olawale O Ajibola[1,2*], Sunday Aribo[1], Oladeji O Ige[1], Ojo J Akinribide[1], Samuel O Akinwamide[1], Peter A Olubambi[1]

[1]Center for Nanoengineering and Tribocorrosion, University of Johannesburg South Africa
[2]Materials and Metallurgical Engineering Department, Federal University of Oye-Ekiti, Nigeria

Corresponding author: olawale.ajibola@fuoye.edu.ng

Abstract. The frictional sliding behaviour of cast Al-Si alloy was investigated with and without DOT 4 hydraulic brake fluid lubrication/environments. Cast Al-Si specimen was produced from wrought alloy and the specimen surface was metallurgically prepared. The as-cast product was characterized with aid of micro-hardness tester, the chemical composition was analysed by X-ray fluorescence (XRF), while X-ray diffraction (XRD) was used to identify phases present, microstructures and surface examinations were done on a high resolution metallurgical Optical Microscope. The properties of the polished and worn sample surfaces were examined by Scanning electron microscope (SEM) equipped with Energy dispersive X-ray (EDS) facilities. A 6mm diameter alumina ball under varying normal loads (3-10 N) was rubbed against the surface of the cast Al-Si coupon specimen. The dry and wet sliding wear behaviour was assessed based on the CoF-sliding time behaviour. From the results, the average CoF values of 0.9064 for dry and 0.2038 for wet contacts were obtained. Comparatively, 1.03E-06 and 4.21E-07 minimum wear intensities; and 467290 and 132170 wear resistances were obtained for dry and wet contacts respectively. Behaviour of cast Al-Si in DOT4 were compared with its performance in DOT3 regarding the hydraulic brake system application.

1. Introduction

Apart from steel, Al-Si cast alloys have been widely employed to produce automotive components working at ambient and fairly high temperature (up to 200 °C) due to excellent characteristics such as low-cost manufacturing, excellent castability, high specific strength and recyclability [1, 2]. Al-Si alloys have a wide range of applications in the automotive and aerospace industries due to an excellent combination of castability and mechanical properties, as well as good corrosion resistance and wear resistivity. Si is the main and most important alloying element of Al-Si cast alloy. Cu and Mg are commonly added to improve the strength at room and elevated temperatures and enable the possibility of heat treatment [3]. The size, morphology and distribution of microstructures control the mechanical and tribological properties of Al-Si alloys [4-6].

In previous reports, the authors focused on the characterisation, usage, wear and corrosion of A6061 aluminium alloys in the DOT3 fluid; to explore and report research findings on the automobile engines study area as applicable to the automobile hydraulic brake cylinder system. The present work is motivated as part of reported large project [7-13] on the study of aluminium alloy in fluid and compared with other materials already reported [14-17]. The present work compares the Al-Si behaviours in the DOT4 with the A6061 alloy in DOT3 previously reported [7-11] whereas, in some previously reported cases, test sample materials were produced from scrap sources [8-9].

2. Experimental procedures

2.1 Materials, characterization and sample preparation

The starting material for the study is wrought aluminium-silicon alloy, received in the form of cylindrical billet (Table 1). The selection of sand-casting variables and adapted procedures complies with the best practices and has been discussed elsewhere [11-13]. The alloy specimen was shaped and cut to 25 mm diameter by 10 mm thickness cylinder size specimen and their surfaces were grinded and polished with 0.2 μm fumed silica suspension to get uniform and smooth mirror-like finishing surface. Five-point micro-indentations were made on the surface of the sample at regular intervals of 2 mm distance apart with the dwell time of 10 s on Vicker's scale HV/1 with a load of 1kgf using the InnovaTest - Falcon 500 model hardness tester.

2.2 Tribological tests

The standard tribometer (Anton Paar, TRB model) was used for the experiments. A 6mm diameter alumina ball was rubbed against the surface of the coupon specimens under varying normal loads. The tribometer was run at sequence count of 10 in a single-way (rotational) mode and at radius also varying from 2.00 to 12.00 mm, and acquisition rate of 10.0 Hz with stop condition of 4.00 m. To assess the behaviour by the sliding frictional/wear (CoFs) method, the normal load was varied from 3, 4, 6, 8 to 10N. The CoF sliding time curves were plotted. The experiments were prepared, performed, analysed and reported considering the ASTM G 115-04, G 40-02; G 99 and G 133, DIN 50324, and ASTM G 99 -95a.

2.3 Microstructural analysis

The chemical composition was analysed by X-Ray Fluorescence (XRF) using the Rigaku ZSX Primus II model. The microstructures and surface examinations were performed by using a high resolution metallurgical optical microscope (Zeiss, MTB 2011 model). The morphologies of the as received and worn sample surfaces were then examined by scanning electron microscope (Tescan Vega 3) equipped with energy dispersive X-ray spectrometer (EDS) facilities. The X-ray diffraction was done to examine phases of the Al-Si alloy material using X-Ray Rigaku Ultima IV.

3. Results and discussion

The combination of the XRF, XRD with the SEM and EDS results (Table 1, and Figure 1) validates the chemical composition, phases and microstructures present in the Al-Si sample as compared with results reported in the previous findings [10,13].

The tensile strength of a material is a function of the hardness of the material [13]. The wear resistance of a material is also a function of the hardness of the material. The harder a material is, the larger its resistance to wear. The present result shows that cast Al-Si test sample has a higher hardness (95.42) value as compared with the 65.7 obtained from the cast A6061 alloy previously reported; it signifies the impact of increased Si content on the higher hardness values in the Al-Si alloy [2] than the A6061 samples reported. Thus, it signifies higher wear resistance potential than the cast A6061 alloy earlier tested in DOT3 [7-8, 10-11].

Table 1. Composition of Al-Si alloy by XRF analysis.

Elements	Al	Cu	Si	Zn	Mn	Fe	Mg	Co	Ti	Sn	Cr	Others
Mass%	74.02	2.01	19.35	1.79	0.36	2.12	0.02	0.01	0.045	0.032	0.042	Bal

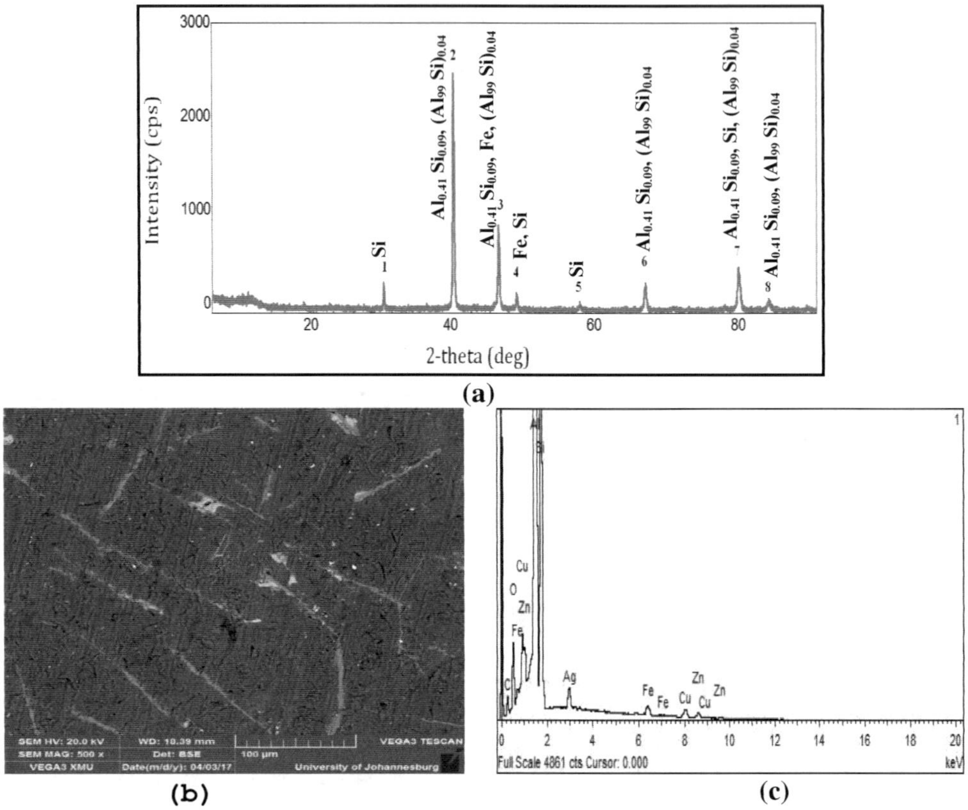

Figure 1: Phase analysis of Al-Si alloy sample by XRD (a), SEM images showing the (b) microstructure, and (c) EDS of Al-Si alloy.

3.1 Wear rates determination by weight loss method

The results of weight losses measured from the dry and wet sliding wear tests performed on the Al-Si samples in DOT4 brake fluid are shown in Figure 2. The testing is compatible with DIN50324 and ASTM. The tests were carried out under varying loads of 3, 4, 6, 8 and 10N. Wear rate increases with the applied normal load for both dry and wet sliding wears, though the wear rate values are much lower in the latter. This agrees with the characteristic behaviour of the cast A6061 alloy in the DOT3 fluid reported previously [12]. The wear rates of the Al-Si specimens subjected to ball sliding actions of the tribometer and measured from weight losses are compared with wear rates calculated from the volume loss and wear rate equations [8]. The wear track width *d* was digitally measured from the high-resolution Zeiss, (MTB 2011 model) optical microscope.

Meanwhile, the trends of the curves obtained from the specimen wear rates (m^2N^{-1}) calculated from the volume loss equations [11] on the similar scale are defined by the equations (1) and (2) for the dry and wet sliding wear characteristics with $R^2 = 0.8781$ and $R^2 = 0.8171$ standard deviations respectively.

$$W_{dc} = 2E\text{-}07e^{0.3747Fn} \quad (1)$$

$$W_{wc} = 5E\text{-}08e^{0.481Fn} \quad (2)$$

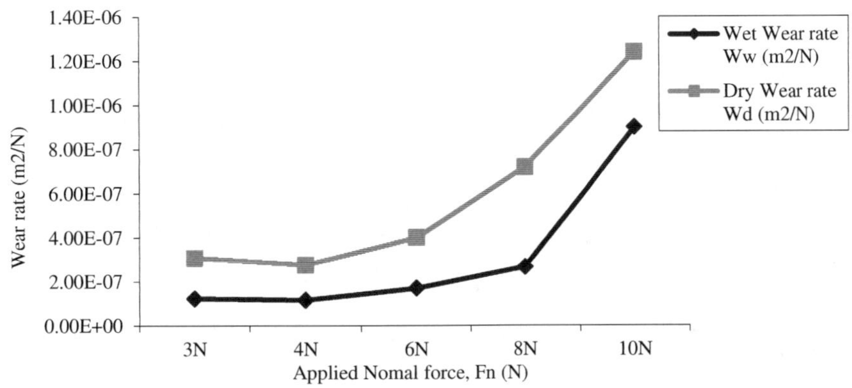

Figure 2: Mass wear rates obtained from the measured weight loss of Al-Si alloy samples in DOT4 brake fluid (wet) and air (dry)

3.2 CoF of dry and wet sliding wear of cast Al-Si sample in DOT4

Table 2 is the summarised data for the CoF for the as-cast alloy in dry and wet environments at various loads. During the dry contact of 3N load, the CoF increased to 1.16 (max) with a mean value of $\mu = 0.858$ obtained at 1910 sec dwelling time of the frictional contact. The wear test was conducted in a wet condition with an application of the same loading was applied to them, thus results are presented in Fig. 2. In this instance, the CoF diminished from 0.627 to obtain $\mu = 0.233$ stability reached after 39.9459 m contact distance (cycle = 2119.36) and 1276 sec sliding time. The increase in the applied load to 4N showed an incremental change in the CoF under the dry contact wear. The μ increased to 1.16 max with a mean value of $\mu = 0.998$ obtained after 959.60 sec sliding time. Meanwhile, this was not so for the wet test in the DOT4 fluid run under 4N load. In this case, there was a reduction in CoF from 0.416 to 0.161 after the 138 sec of contact. A mean $\mu = 0.193$ value was obtained after 766.92 sec sliding time. With further increase in normal load to 6N also influenced CoF value from 1.29 (max) to a mean value of $\mu = 0.823$ after the 641.66 sec total sliding time. The oil reduced the CoF to 0.517 after 546.13 sec and 39.9581 m sliding distance as compared with dry wear. The characteristic of wet wear is different in that, the CoF reached the maximum value of 0.811.

For the period of the dry contact of 8N applied load the μ increased to 1.49 max and a mean value of $\mu = 1.03$ was obtained at 408.22 sec frictional contact time. Comparatively, the CoF fluctuates between 0.423 (6.98 sec) and 0.161 (180 sec) and the stabilised during the wet wear. In this case, the CoF reduced from 0.423 to obtain CoF value of $\mu = 0.233$ stability obtained after 39.94 m contact distance (cycle = 706.56) and 426.30 sec sliding time.

The 10N applied load showed an incremental change in the CoF under the dry contact wear. The CoF initially increased to 1.49 max and with a stable CoF value of 0.823 obtained after 366.4 sec sliding time and 605 wear cycle. On the other hand, for the wet wear test in the DOT4 fluid run under 10N load, the CoF becomes stable at $\mu = 0.163$. Moreover, a mean CoF of 0.193 was obtained at the 347.94 sec total sliding time. The whole results reflect slight similarities in the CoF – time behaviour of Al-Si alloys subjected to a varying small range (of 5 to 9 N) frictional force reported previously by Mathavan and Patnaik [14].

Table 2. Wear parameters of cast Al-Si alloy under dry and wet.

Normal load (Fn)	***3N***	***4N***	***6N***	***8N***	***10N***
CoF (dry)	0.858	0.998	0.823	1.03	0.823
CoF (wet)	0.233	0.193	0.517	0.233	0.193

Dry track width (mm)	0.95	1.13	1.31	1.48	1.74
Wet track width (mm)	0.34	0.35	0.47	0.48	0.62
Intensity (dry)	1.03E-06	1.04E-06	1.95E-06	4.18E-06	7.57E-06
Intensity (Wet)	5.80E-07	4.21E-07	4.93E-07	7.85E-07	2.14E-06
Resistance (dry)	967118	958772	512820	239120	132170
Resistance (wet)	3611904	2375296	2029632	1274534	467290

3.3 Optical microscopy and SEM of wear tracks

Figure 3a shows the optical micrograph of the microstructure of the polished cast Al-Si alloy surface (control) prior to the dry sliding and wet sliding wear tests. The optical micrographs of the wear scar of cast Al-Si alloy surfaces under the dry sliding and wet sliding in DOT4 oil impressed with different applied loads (3 and 10N) are presented in Figure 3(b)- 3(e). In all instances of impressed loading, observations show that wear is more severe on the dry surfaces that the oil lubricated interfaces. For more clarity, the SEM images showing wear scar patterns of cast Al-Si alloy under (a) dry and wet sliding in DOT4 oil under 2N applied load are presented in Figures 4a and 4b. The nature of wear defined on this system is sliding wear and comparatively, the volume of materials that were removed from the dry and wet surfaces is considerably large across all varied loads for the specimen in both different (dry and wet) conditions. The asperities observed (Figures 3 and 4) for dry and wet shows direct proportion to the load increase. However, the wet grinding has a lower wear loss when compared to the dry grinding at the same loads.

The SEM images reveal conspicuous wear tracks for both dry and wet across the different loads as shown in the image for 2N loading (Figure 4(a) and (b)). This incidence can be explained in diverse ways which include the consideration for the state of specimen surface finish [18]. It is renowned that lesser wear generally occurs when a rough surface slides against a smooth surface and greater wear when both surfaces are either extremely rough or very smooth because of increased adhesive or abrasive wear constituents correspondingly. Since the test specimen surfaces were prepared by grinding from 3.0 µm to 0.2 µm finishing grit, the Al-Si alloy surfaces in contact with a hard alumina ball experienced sliding wear relative to the load. Table 2 contains the wear parameters and frictional wear data generated from the tribometer with intensity and wear-resistance calculated based on wear equations for all specimens [19].

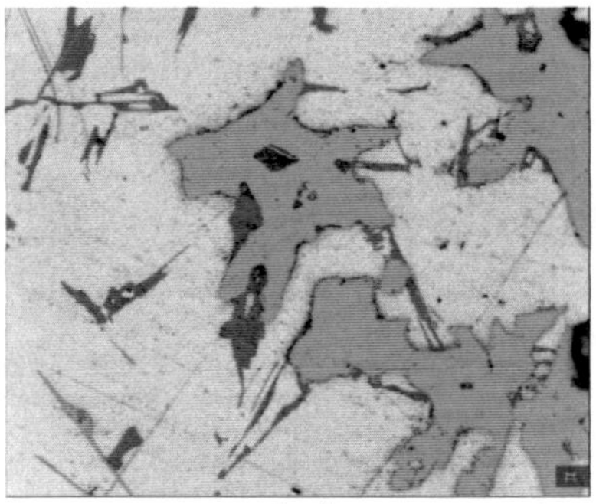

(a)

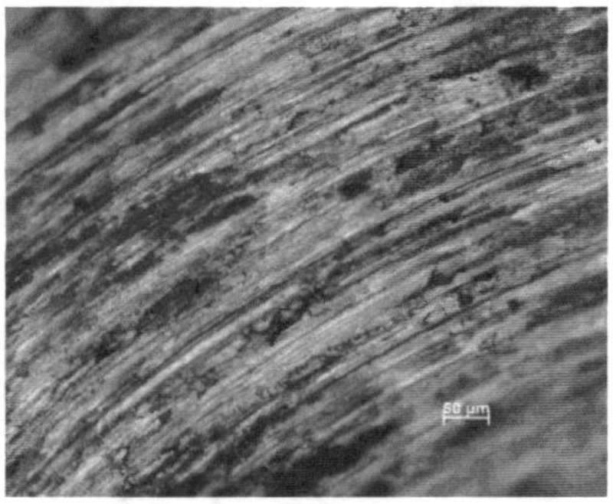

(b)

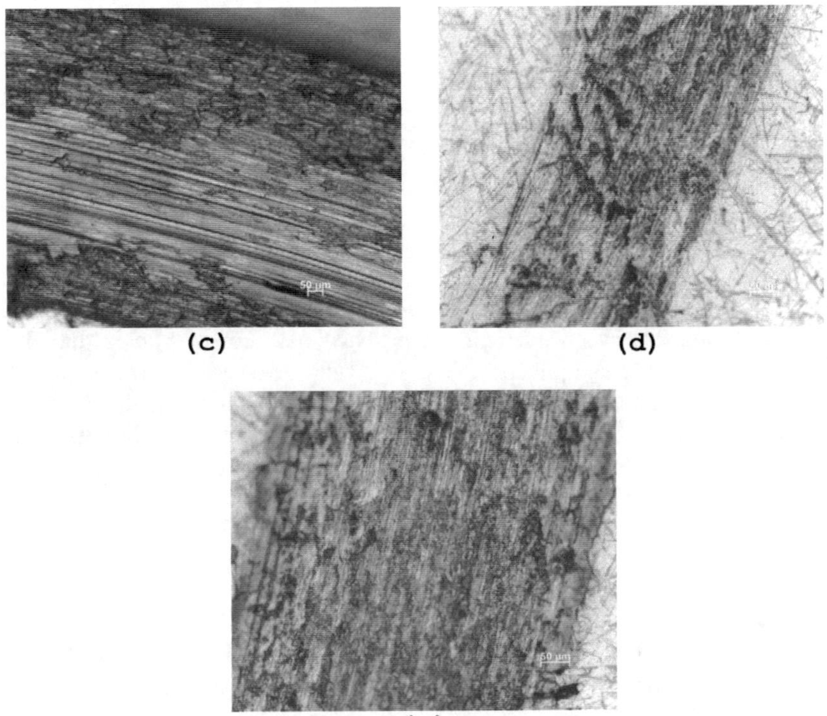

Figure 3: Optical micrographs of surfaces (a) before and wear scar patterns after dry sliding with (b) 3N , (c) 10 N and wet sliding with (d) 3 N and (e) 10 N loads.

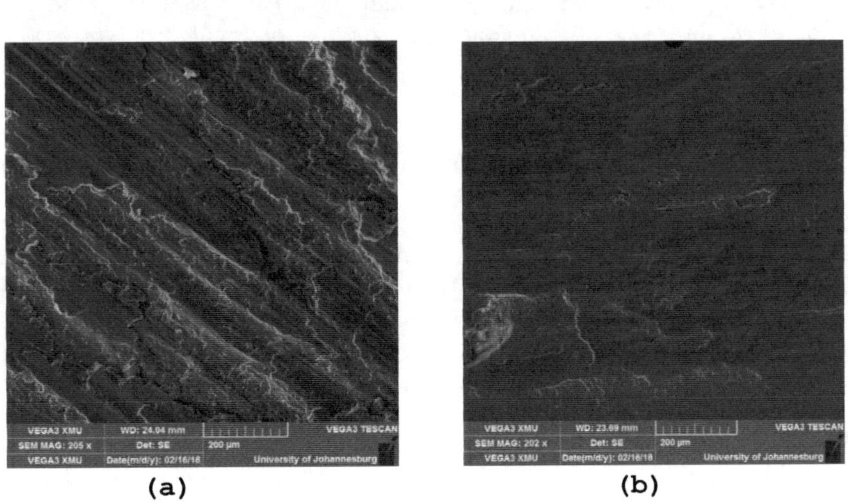

Figure 4: SEM images showing wear scar patterns of cast Al-Si alloy under (a) dry and (b) wet sliding in DOT4 oil under 2N load.

The trends of some other wear parameters (wear track width, wear intensity, and wear resistance) with normal force variation (3, 4, 6, 8 and 10N) in the case of cast Al-Si alloy subjected to dry and wet sliding in the DOT4 fluid are compared and presented in Table 2. It shows the wear track width variation with applied load. The wear track width increased with increasing applied wear loads for both dry and wet sliding wear experiments even-though the values obtained are smaller as in the case of the wet sliding wear. It also shows that the impact of wear intensity is higher on the dry sliding that the wet sliding of the ball on the Al-Si alloy. The resistance is the hindrance to wear and it is measured by the inverse of wear intensity. Also, the wear resistance had been simply defined by Ludema, [16] as the reciprocal of wear volume (in cm^3). The wear resistance reduces with increase in applied load values for both dry and wet sliding conditions. In the present case, the wear resistance of the test cast Al-Si sample is comparatively higher in DOT 4 fluid than for the previously reported cast A6061 samples tested in DOT3 [11, 20]. It was reported in previous studies that Al-Si alloys wear resistance improves with the increased in Si content and as well as load-bearing capacity [17].

4. Conclusion

This study carried out a comparative analysis of the wear characteristics of as-cast Al-Si alloy in DOT 4 fluid using both dry and wet sliding procedures. Data are obtained for the relationships of mass wear, wear rate, wear intensity and wear-resistance. The basic conclusion is that increased normal force loading of aluminium alloy sample leads to a significant decrease of the wear-resistance in both dry and wet sliding contacts. The present study reports a higher wear resistance of cast AlSi sample in DOT 4 fluid than the previously reported cast A6061 tested in DOT3. Also, higher values of CoF were observed in AlSi alloy as the load increases while lower wear rates characterised the AlSi alloy in DOT4 with and without lubrication than the A6061 wear in DOT3 previously reported. A vista of study for future consideration of the cast Al-Si alloy in DOT 4 fluid, having in view their operation under various exploitation conditions of interaction that is in a multiphase system.

5. References

[1] Akinwamide, S.O., Lemika, S.M. and Abiodun, B., 2019. Study of Microstructural and Mechanical Properties of Stir Cast Al (SiC-Mg-TiFe) Composite. *Fluid Dynamics & Materials Processing*, *15*(1), pp.15-26.

[2] Islam, M.A. and Farhat, Z.N., 2011. Effect of porosity on dry sliding wear of Al–Si alloys. *Tribology International*, *44*(4), pp.498-504.

[3] Stadler, F., Antrekowitsch, H., Fragner, W., Kaufmann, H., Pinatel, E.R. and Uggowitzer, P.J., 2013. The effect of main alloying elements on the physical properties of Al–Si foundry alloys. *Materials Science and Engineering: A*, *560*, pp.481-491.

[4] Olofsson, J., Svensson, I.L., Lava, P. and Debruyne, D., 2014. Characterisation and investigation of local variations in mechanical behaviour in cast aluminium using gradient solidification, Digital Image Correlation and finite element simulation. *Materials & Design (1980-2015)*, *56*, pp.755-762.

[5] Andoko, F., Puspitasari, P. and Gapsari, F., 2018. The analysis of strength and fracture morphology of Al-Si compound made from moulding sand formulation with bentonite binding material and Portland cement. *Journal of Achievements in Materials and Manufacturing Engineering*, *89*(1).

[6] Stojanovic, B., Babic, M., Mitrovic, S., Vencl, A., Miloradovic, N. and Pantic, M., 2013. Tribological characteristics of aluminium hybrid composites reinforced with silicon carbide and graphite: A review. *Journal of the Balkan Tribological Association*, *19*(1), pp.83-96.

[7] Ajibola O.O, Oloruntoba D. T., Adewuyi B.O., 2015. "Design and performance evaluation of wear test jig for aluminium alloy substrate in hydraulic fluid". *African Corrosion Journal*, 1(1), pp 40-45.

[8] Ajibola O O., Ige O O.,. Olubambi P A (2018), Wear and Corrosion of Wrought A6061 Aluminium Alloy in DOT3 Brake Fluid. *International Journal of Engineering & Technology*, 7(2) pp512-519.

[9] Ajibola O.O, Oloruntoba D. T., Adewuyi B.O., 2016 "Investigation of corrosion of cast aluminium alloy piston in brake fluid". *African Corrosion Journal*, 2(2) pp 1-11.

IOP Conf. Series: Materials Science and Engineering **628** (2019) 012012 doi:10.1088/1757-899X/628/1/012012

[10] Ajibola O.O, Oloruntoba D.T, and Adewuyi B.O, 2014. Metallurgical study of cast aluminium alloy used in hydraulic master brake calliper. *International Journal of Innovation and Scientific Research*. 8(2):324-333.

[11] Ajibola O.O, Adewuyi B.O, and Oloruntoba D.T, 2014. Wear behaviour of sand cast eutectic Al-Si alloy in hydraulic brake fluid. *International Journal of Innovation and Applied Studies*, *6*(3), p.420-430

[12] Ajibola O.O, and Oloruntoba D.T., 2015, Wear and corrosion of cast Al alloy piston with and without brake oil. *Indian Journal of Materials Science*, Article ID 763618, p10.

[13] Ajibola O.O, Oloruntoba D.T, and Adewuyi B.O, 2015, Effects of moulding sand permeability and pouring temperatures on properties of cast 6061 Aluminium alloy. *International Journal of Metals*, Article ID 632021, p13.

[14] Mathavan J.J and Patnaik, A. 2016, *Mater. Sc. Eng*. 149, 2016

[15] Sahin, Y. and Özdin, K., 2008. A model for the abrasive wear behaviour of aluminium based composites. *Materials & Design*, *29*(3), pp.728-733.

[16] Ludema, K.C., 1992. Sliding and adhesive wear. *ASM handbook*, *18*, pp.236-241.

[17] Torabian, H., Pathak, J.P. and Tiwari, S.N., 1994. Wear characteristics of Al-Si alloys. *Wear*, *172*(1), pp.49-58.

[18] Garzino – Demo G.A, Lama F.L, 1995, Friction and wear of metallic and non – metallic surfaces, *Surface and Coatings Technology*, Vol 76–77, Part 2, 487-493, 1995

[19] Ajibola O.O, Oloruntoba D. T., 2015, Effect of MgFeSi inoculants on properties of Cast 6061 Al Alloy for brake master piston application, *Indian Journal of Materials Science,* Article ID 756219, p10.

[20] Anton-Paar G74IB001EN-B TRB User Manual, Tribometer Software Version 7, pp1-141, www.anton-paar.com

Acknowledgement
The author would like to acknowledge the Management of Premier Wings Engineering Services, Ado Ekiti, Nigeria for providing the materials and some workshop services used for this work.

IOP Conf. Series: Materials Science and Engineering **628** (2019) 012013 doi:10.1088/1757-899X/628/1/012013

Study of photovoltaic solar materials development trend by deductive logic method using big data

Yi-Huei Su[1] and Jung-Hua Wu[1]

[1]Department of Resources Engineering, National Cheng Kung University, Tainan, 701, Taiwan.

Email: hwaa@mail.ncku.edu.tw

Abstract. In recent years, the development of solar materials in the field of science and technology has experienced many major breakthroughs. This study collects relevant data to evaluate the development of solar energy technology based on deductive reasoning approach in the popular types, and then utilize big data to investigate the trend of smart materials development in solar energy research. This study aims to understand the impact of materials science on the development of photovoltaic solar energy and understand the trends in the research field from the number of published journals in past years.

The period of study in this paper is from 1970 to 2018. The research scope is limited to an analysis of solar journals from Web of Science®, which indicates that various solar materials have very obvious difference in growth trends.

From 1970 to 2010, although the United States was the country with the largest number of published solar journals, the percentage gradually decreased. Then from 2011 to 2018, China became the first among the top countries in solar energy materials publications.Among them, the field of "Dye- sensitized solar" grew the most, the highest research publications in the field of solar energy research is "Materials science".

1. Introduction

In recent years, photovoltaic solar energy development has been recognized as an emerging clean and environmental friendly technology. The evolution of the development of solar energy research and solar technology has led to a reduction in power and energy supply uncertainties, industrial development, and a reduction in carbon emissions.

At present, the most popular types of solar materials are "solar material", "photovoltaic material", "Perovskite solar cell", "Copper Indium Gallium Diselenide" (referred to as CIGS), "Dye-sensitized solar", "GaAs solar cell" "Silicon solar cell," and "Thin film solar cell" based on Web of Science®.

In terms of the research field of solar materials, popular research topics relative to solar energy are "Astronomy & Astrophysics", "Geochemistry geophysics", "Physics", "Engineering", "Energy Fuels", "Science and Technology other topics", "Geology", "Materials science", "Meteorology atmospheric sciences", "Chemistry", "Thermodynamics", "Optics," and "Electrochemistry". Therefore, these comparisons among 1970-2018 annual reports are investigated.

The paper mainly discusses about the solar energy materials during 1970-2018, using deductive reasoning to analyze big data operations, speculation trends, and the technology development of smart materials in the research field of solar materials, research fields, and publications of national journals.

IOP Conf. Series: Materials Science and Engineering **628** (2019) 012013 doi:10.1088/1757-899X/628/1/012013

2. Data collection and analysis

2.1 Approach used.

This methodology is the deductive reasoning, which includes one or more descriptions, analysis logic, and inductive reasoning conclusions[1].

The rules of logical deduction affirms the consequent and rules of separation as Modus ponens Description includes the affirming of the consequent (P->Q) and the proof of rules of separation (P). Modus ponens conclusions are obtained by the conditional classification, as a conclusion (Q). The parameters are described as the following list:

1. P->Q (Conditional statement and description)
2. P (Antecedent description)
3. Q (Inductive results)

In this deductive reasoning method, the consequent (Q) is obtained as the conditional statement, description (P->Q) and Antecedent description(P).

2.2 Data collection and analysis.

Our data was collected from the relative publications about solar materials from Web of Science®. The collected data are intended to explore the stages of the research fields of solar materials about solar technology in the range between 1970 and 2018 by running the software Statistical Product and Service Solutions (SPSS), IBM®.

3. Results and discussion

The evaluation of solar journals can be divided into five processes, namely, Early Development of Solar energy, Energy Crisis Push Improvements of Solar Energy Systems, Solar Energy Meets Carbon Emission Reduction, Claim Solar Energy in Reducing Carbon Emission, and Recent Development of Solar Energy Applications (see Figure 1.).

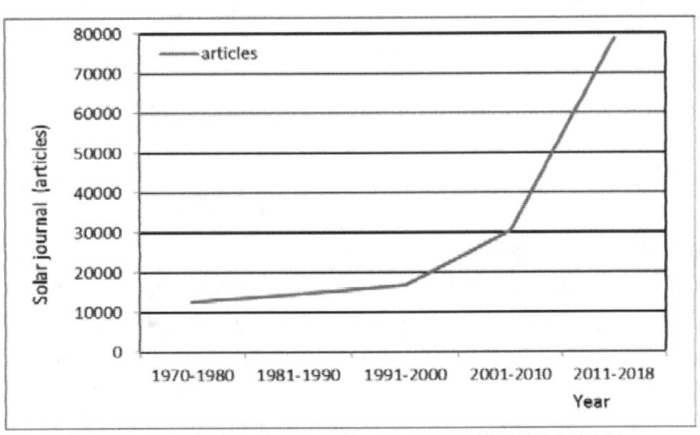

Figure 1. Solar journal processes(1970-2018).

In the 1970s, because of the energy crisis, many countries have increased their willingness to develop solar energy and reduce carbon emissions at the same time.

In 1990, there were many studies suggesting that if the global population continued to increase, per capita greenhouse gas emissions would also increase rapidly. In the period between 2001 and 2010, many scholars pointed out economic and climate change issues related to solar energy, including determining how to improve the efficiency of solar energy due to the continued increase in the solar energy market and the short supply of polysilicon. In the rural areas of developing countries, solar energy will effectively improve the electricity supply in remote areas.

From 2011 to 2018, as solar power technology matured, the technical start from basic science was used to solve application problems and improve conversion efficiency. Furthermore, developing an understanding of the different regions of the world is still a very important issue, because of the uncertainty related to energy use on a commercial basis since it is affected by the economy, public attitudes, and social technical analysis.

Currently, solar energy technology is developing, and there are many ways to foster and improve the conversion efficiency of solar energy. These include solar spectrum research and the use of ultra-violet light, visible light, infrared light, and different light properties for different uses in solar power systems [2], the use of light concentration[3] (see Figure. 2.), plasmonics effect [4-5], Foster energy transfer[6-7], nanoantenna [8], high conductivity materials[9], quantum energy transport [10-16], photon entropy in concentrator and angularly restricted systems[17], textured screen-printed solar cells that exceed the limit of plasmonic light trapping [18], a priori calculation of the internal reflectance of a solar cells[19], Nanofluid-based optical filter optimization [20], Plasmonic photocatalyst[21], and so on [22]. Recent development of solar energy applications [23-29] and uses of solar energy have improved [30-33].

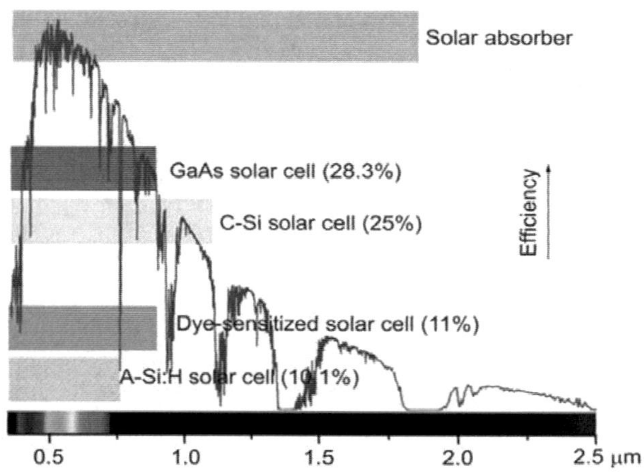

Figure 2. Solar cell efficiency.

We report some state-of-the-art technologies, including photon entropy in concentrator and angularly restricted systems, exceeding the limit of plasmonic light trapping, the internal reflectance of solar cell, plasmon-sensitized solar cell, Nanofluid-based optical filter optimization, Plasmonic photocatalyst. These technologies have potential applications in the field of solar energy device and also enhance the conversion efficiency of solar cells [34-39].

From the emergence of Solar energy technology in the 1970s to the rapid growth of Solar energy technology in the present day, understanding the trend of the development of Photovoltaic Solar Materials then becomes important. The following are the results of the research: Comparison of solar materials; the trend of the publications about solar materials in different research area during 1970-1980, 1981-1990, 1991-2000, and 2001-2010; the recent trend of the publications about solar materials in total during 1970-2018 in percentage; top ten countries publishing in the field of solar materials during 1970-1980,1981-1990, 1991-2000, and 2001-2010, and 2011-2018; Solar energy materials researches published by different countries during 1970-2018 in percentage.

The number of publications included in the top research field, including solar material, photovoltaic material, perovskite solar cell, CIGS, dye-sensitized solar cell, GaAs solar cell, silicon solar cells, and thin film solar cells, during the 1970-2018 are presented in Figure. 3.

Figure. 3 shows that the trend of the publications in "solar material" is flat from 1970 to 2010, and grew slightly during 2011-2018. The trend of the publications in "photovoltaic material" increases slowly during 1970-2018; the trend of the publications in "perovskite solar cell" is presented from 1970-2010. The number of publications related to "perovskite solar cell" grows at a high rate from 2011-2018; the trend of the publications in "CIGS" grows slowly in 1970-2018; the trend of the publications in "Dye-sensitized solar cell" does not change significantly during 1970-2000, but continues to grow from

2001-2010, and grows at a high rate in 2011-2018. The trend of the publications in "GaAs solar cell" does not grow significantly from 1970-2018. For the "Silicon solar cell" publications, the growth is flat from 1970-2000, and grows slightly from 2001-2010 and 2011-2018; the trend of the publications in "thin film solar cell" shows small growth from 2001-2010 and 2011-2018.

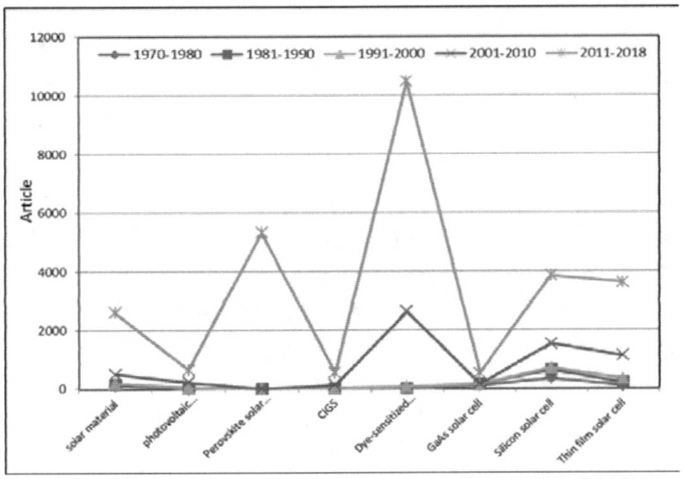

Figure 3. Comparison of solar materials (1970-2018).

In 2011-2018, the biggest increase is observed in "Dye-sensitized solar cell", which is followed by "Perovskite solar cell". The smallest growth was observed for "Silicon solar cell" and "Thin film solar cell" in 2001-2018.

The top ten research areas related to solar materials based on the number of publications during 1970-2010 and the recent 8 years are shown in Figure.4 and Figure. 8, respectively.

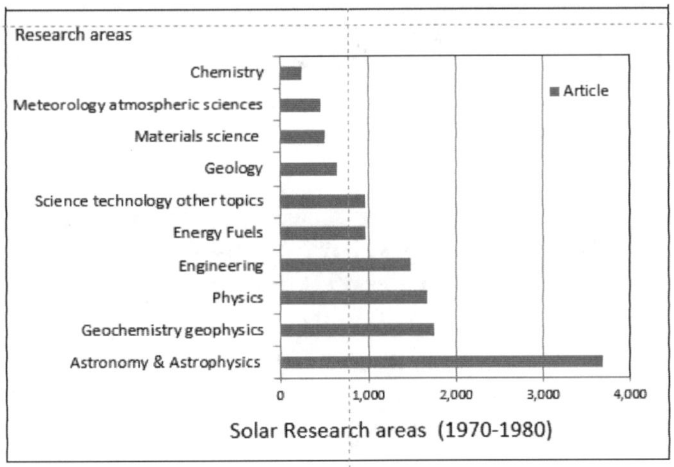

Figure 4. The trend of the publications about solar materials in different research area during 1970-1980.

From 1970 to 1980, the top three solar energy research fields were "Astronomy Astrophysics", "Geochemistry geophysics" and "Physics". The top 8th, 9th, and 10th ranks were "Materials science", "Meteorology atmospheric sciences" and "Chemistry", respectively.

From 1981 to 1990, the top three solar energy research fields were "Astronomy Astrophysics", "Energy Fuels," and "Physics". The top 8th, 9th, and 10th ranks were "Geochemistry geophysics", "Thermodynamics," and "Chemistry". Moreover, petrochemical energy and engine field play an important role during those ten years. From 1991 to 2000, the top three solar energy research fields were still "Astronomy Astrophysics", "Physics," and "Energy Fuels". "Geochemistry geophysics", and "Thermodynamics" drops to the 9th and 10th rank respectively, and a new field, "Science Technology and Other Topics", drops to the 8th rank from the 5th rank.

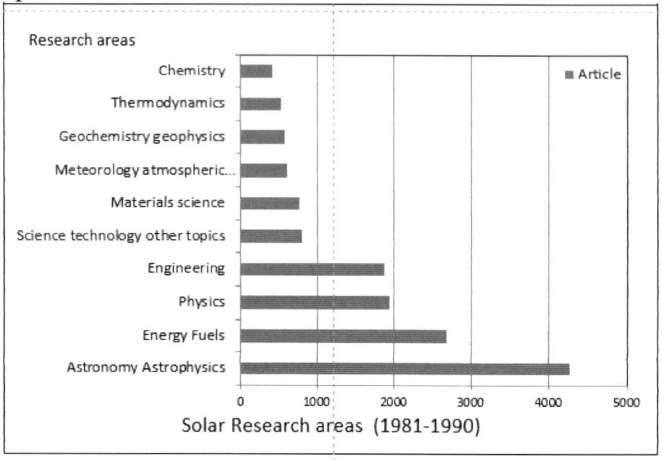

Figure 5. The trend of the publications about solar materials in different research area during 1981-1990.

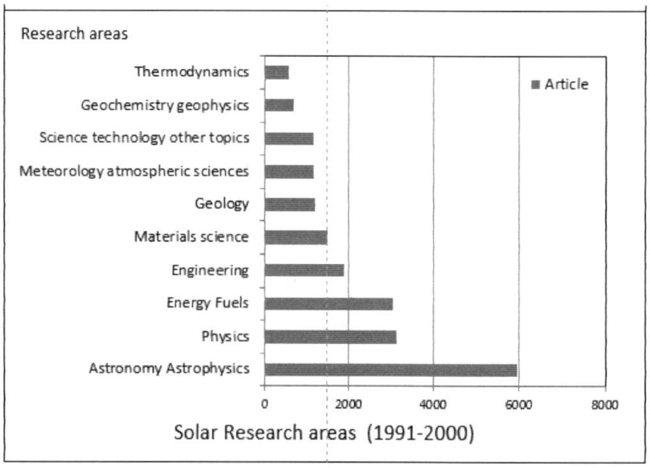

Figure 6. The trend of the publications about solar materials in different research area during 1991-2000.

From 2001 to 2010, the top two solar energy research fields remained to be "Astronomy Astrophysics", and "Physics", while "Materials Science" takes the top third spot from 5th rank in the last 10 years. The top 8th , 9th , 10th rank were "Meteorology atmospheric sciences", "Geology" and "Geochemistry geophysics", respectively.

In 2011-2018, however, "Materials science" takes the top rank, followed by "Physics", and "Chemistry". The top 8th, 9th , and 10th rank were "Thermodynamics", "Optics," and "Electrochemistry", respectively. The results show that from 1970-2018, "Science technology other topics" and "Energy Fuels" consistently appear in the top ten of the field of solar energy research. However, from 1970 to 2000, the top three research fields were mostly principle classes. In 2001-2018, the top three research fields are mostly application classes. In 2011-2018, the first name in the field of solar energy research was "Materials science", and "Optics" and "Electrochemistry" was included in the 9th and 10th rank.

Figure. 9 shows the recent trend of the publications about solar materials in total during 1970-2018 in percentage.

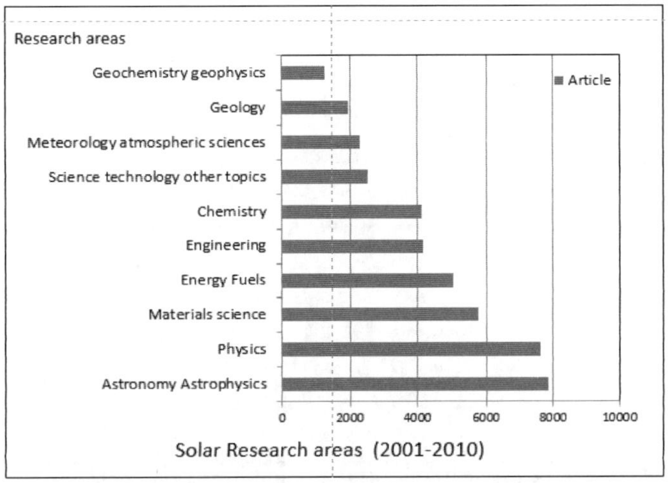

Figure 7. The trend of the publications about solar materials in different research area during 2001-2010.

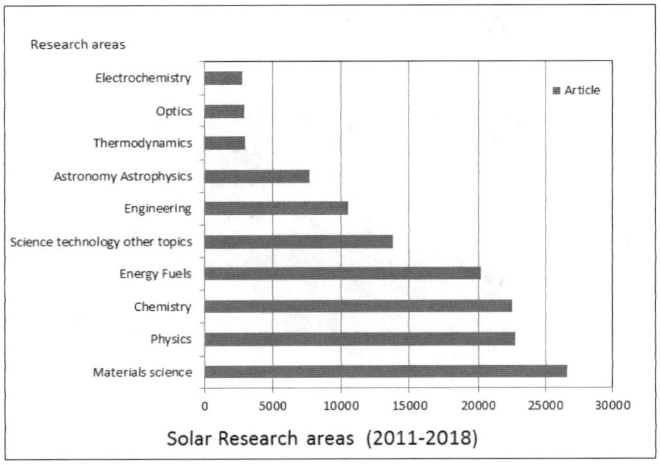

Figure 8. The recent trend of the publications about solar materials in different research area (2011-2018).

As presented in Figure. 9, the top three research fields based on the number of published solar papers are "Physics", "Materials science" and "Energy fuels". In addition, the top 8th, 9th, 10th rank are "Meteorology atmospheric sciences", "Geochemistry geophysics", and "Geology", respectively.
Figure. 10 and Figure. 14 present top ten countries publishing in the field of solar materials, which is divided into five periods by 1970-1980, 1981-1990, 1991-2000, 2001-2010, and 2011-2018.

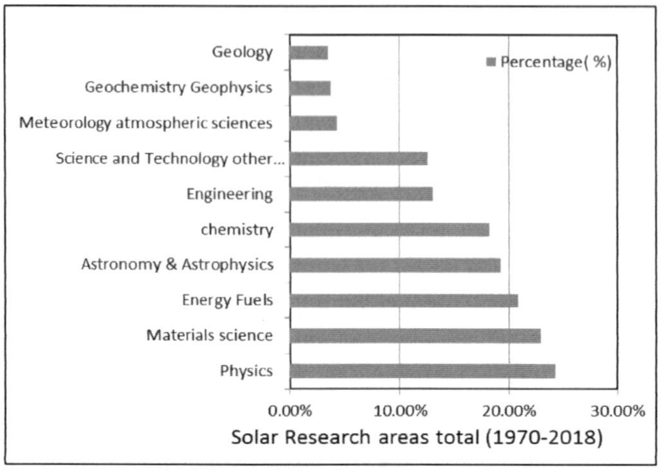

Figure 9. The recent trend of the publications about solar materials in total during 1970-2018 in percentage.

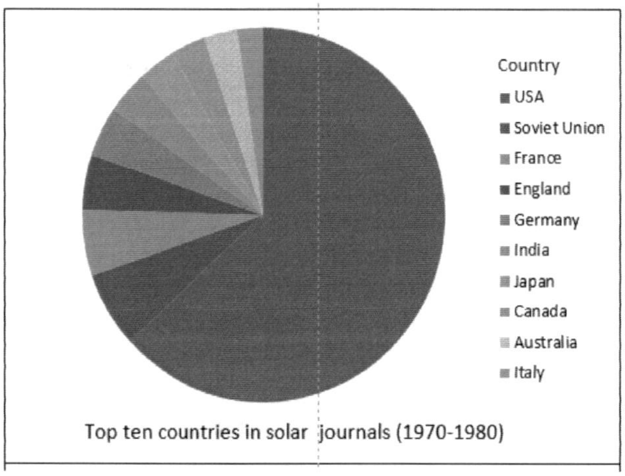

Figure 10. Top ten countries publishing in the field of solar materials during 1970-1980.

Figure.10-13 shows that the top three countries publishing works related to solar materials from 1970-1980 are USA, the Soviet Union, and France. The top 8th, 9th, and 10th rank belong to Canada, Australia, and Italy, respectively. In 1981-1990, the top three are USA, India, and the Soviet Union. The top 8th, 9th , and 10th rank are Australia, Italy, and Canada. From 1991 to 2000, the top three countries publishing in the field of solar materials are USA, Germany and Russia. The 8th ,9th , and 10th rank are Italy, Switzerland, and Australia, respectively. From 2001 to 2010, the top three countries publishing in the field of solar materials are USA, Germany and China, and the 8th , 9th , and 10th rank are Italy, Spain and South Korea. Figure. 14 shows that, in 2011-2018, the top three countries publishing in the field of solar materials are China, USA, and South Korea. The 8th , 9th , and 10th rank are Spain, Taiwan and France.

From 1970 to 2010, although the United States was the country with the largest number of published solar researches, the percentage gradually decreased until China eventually took the spot in the past 10 years. From 1970 to 1990, USA and the Soviet Union are constantly in the top three. From 1991 to 2010, Germany ranked second in solar energy publications. Before 2001, China and South Korea were not listed in the top 10 countries with a significant number of publications in solar materials. However, from 2001 to 2010, China ranked third in the solar materials publications category, and then South Korea became the 10th in the number of solar materials related publications. During 2011-2018, China has become the number one country publishing in the field of solar materials. Moreover, before 2011, Taiwan was not in the top ten countries in solar materials publications. In 2011-2018, Taiwan ranked 9th in the publication field of solar materials.

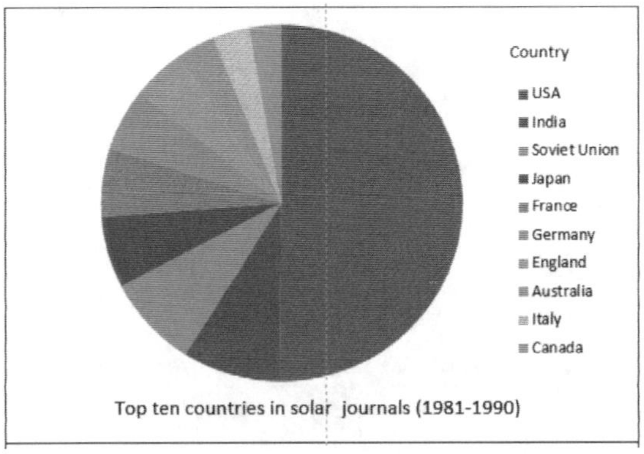

Figure 11. Top ten countries publishing in the field of solar materials during 1981-1990.

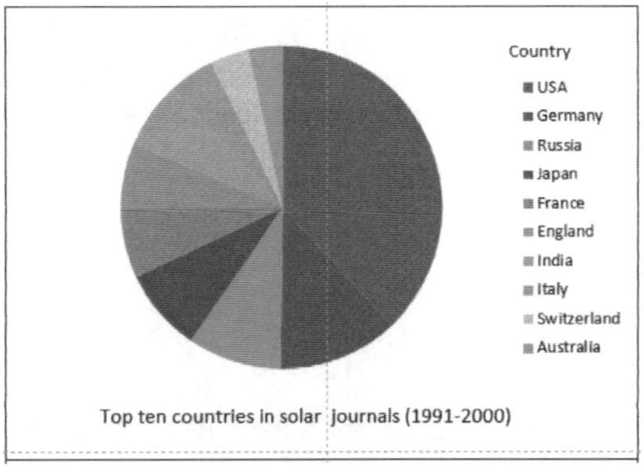

Figure 12. Top ten countries publishing in the field of solar materials during 1991-2000.

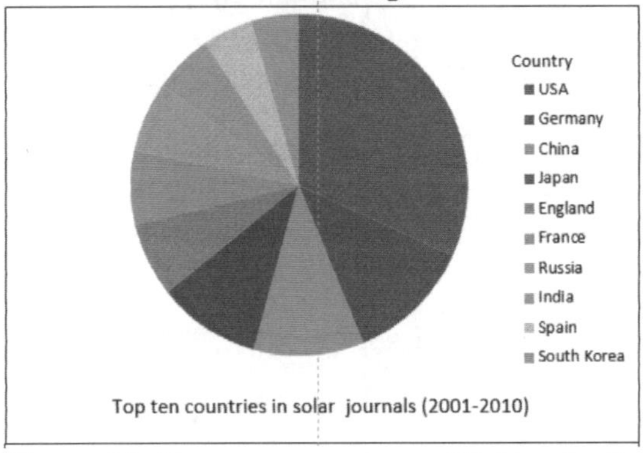

Figure 13. Top ten countries publishing in the field of solar materials during 2001-2010.

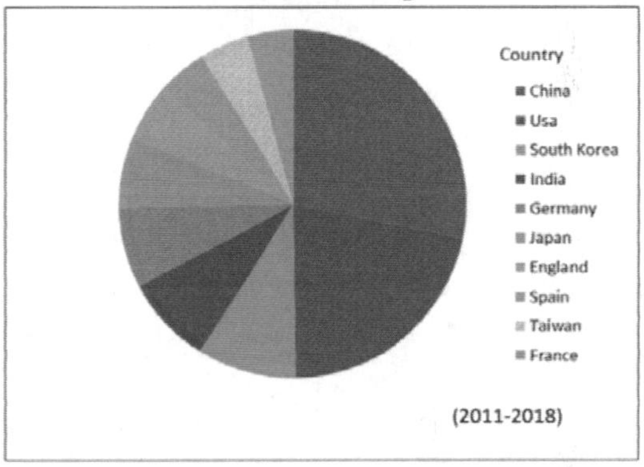

Figure 14. Top ten countries publishing in the field of solar materials in recent (2011-2018).

Combining the above results, for example in United States, from 1971 to 1980, in Web of Science®[40], the number of the research papers on solar materials in the US accounted for 54% of the total amount of journal articles . At the present time, the development of solar technology in the US is more mature than it is in other countries, from 1981 to 2010, compared with previous years, the number of papers on this topic decreased. However, the number of research papers on solar technology in the US was still greater than other countries. From 2011 to 2018, the number of research papers on solar materials in the US accounted for only 26.96% of the total amount of journal articles, and China moved up to first place globally, with US ranking 2nd.

On the other hand, it is worth mentioning that based on the number of the published research papers, China produced one-eleventh of what US produced between 1991 and 2000. In fact, from 1991-2000, the number of solar journals published in China did not even reach ten. In 2004, the amount of published papers from China was a quarter of that produced in the US. Even though China's solar energy journals were published in less than 10 publications in the period from 1991-2000, China has made rapid progress during this period. As a result, China ranked third in the number of solar energy journals published in the following 10 years (2001-2010) and eventually taking the 1st rank from 2011 to 2018. Figure.15 shows the researches of solar energy materials published by different countries in percentage during 1970-2018.

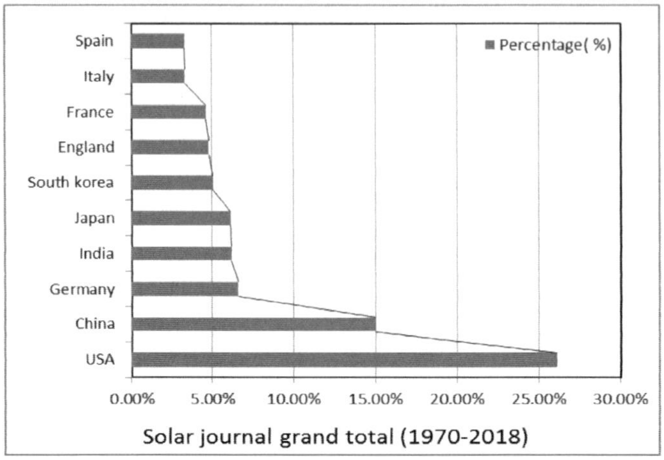

Figure 15. Solar energy materials researches published by different countries duiring1970-2018 in percentage.

The top 10 countries in percentage during 1970-2018 is also shown. As presented in Figure.15, from 1970 to 2018, USA is still the country with the largest number of publications, followed by China and Germany. The top 10 countires are USA, China, Germany, India, Japan, South Korea, England, France, Italy and Spain.

The cumulative amount of the publications about solar materials in national countries from 2011 to 2018 is as follows.

Figure.16 shows the first place is the United States, the second place is China, and the third place is Germany in the field of research.

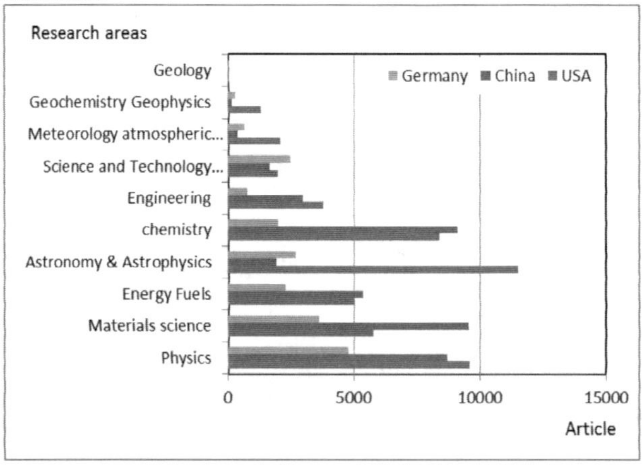

Figure 16. Solar energy materials researches published by different countries comparison duiring1970-2018.

In the beginning of the study, since the ranking of materials science publications are followed, the cumulative publication volume is relatively small. During 2011-2018, the number of materials science publications is ranked 1st. For the cumulative publication volume of journals from 1970 to 2018, the publication volume of materials science is ranked 2nd. Research shows that the cumulative publishing volume of materials science has increased significantly and in recent years, the impact of materials science on photovoltaic solar energy is obvious and significant.

4. Summary

Solar technology has gradually improved over the past decades, and after 2005, the trend of rapid growth was clearly observed. From 1970 to 2018, various solar materials have different growth trends.

Among them, in 2011-2018, "Dye-sensitized solar cell" grew the most, followed by "Perovskite solar cell". The growth trend is flat with photovoltaic material, GaAs solar cell, and CIGS.

In solar research areas, the results show that from 1970 to 2000, the top three research fields were mostly principle classes. In 2001-2018, the top three research fields are mostly application classes. From 2011 to 2018, the top research classification in the field of solar materials is "Materials science". From 1970 to 2010, although the United States was the country with the largest number of published solar materials, the percentage gradually decreased. Finally from 2011-2018, China became the country with the most solar energy materials publications.

In summary, the authors have discovered that the rise of materials science does affect the development of photovoltaic solar energy. Moreover, the rapid increase in the number of published journals in recent years, among them, the cumulative amount of journals in the field of materials science has increased the most. The research also showed that there was an increase in the energy (soft power) of materials science. However, the success of solar energy development still requires the mutual cooperation and development of the industry and the government. It is recommended that future research can be added to patents (industry output results) and R&D expenditure (government investment) be increased, both of which when used in cooperation can help better understand the development trend of solar energy.

5. References

[1] R. J. Sternberg, Cognitive Psychology. Belmont, CA: Wadsworth,6th ed.2010,ch12,pp. 507-510. ISBN 978-1-111-34476-4.

[2] C. Bertram, G. Luderer, and R. C. Pietzcker, "Complementing carbon prices with technology policies to keep climate targets within reach," *Nature climate change* , vol. 5, pp. 235-239, February 2015.

[3] C. F. Guo, T. Sun, F. Cao, Q. Liu , and Z. Ren, "Metallic nanostructures for light trapping in energy-harvesting devices,"*Light: Science & Applications* , vol. 3, pp.e161, April 2015.

[4] M. Memarian, and G. V. Eleftheriades, "Light concentration using hetero-junctions of anisotropic low permittivity metamaterials," *Light: Science & Applications* , vol. 2, pp.e114, November 2013.

[5] Y. H. Su, and W. L. Wang, "Surface plasmon resonance of Au-Cu bimetallic nanoparticles predicted by a quasi-chemical model, "*Nanoscale Research Letters* , vol. 8, pp.408, October 2013.

[6] K. B. Lin, and Y. H. Su, "Photoluminescence of Cu:ZnS, Ag:ZnS, and Au:ZnS Nanoparticles Applied in Bio-LED," *Applied Physics B-Lasers and Optics* , vol. 113, pp. 351-359, May 2013.

[7] K. Sugioka, and Y. Cheng, "Ultrafast lasersreliable tools for advanced materials processing," *Light: Science & Applications*, vol. 3, pp. e149, April 2014.

[8] Y. H. Su, and W. Y. Chen, " Nanoantenna-like Properties of Sea-urchin Shaped ZnO as a Nanolight Filter," *Applied Physics Letters* , vol. 101, pp.101-133, April 2012.

[9] Y. H. Su, Y. K. Wu, S. L. Tu, and S. J. Chang, "Electrostatic Studies of π-π Interaction for Benzene Stacking on a Graphene Layer," *Applied Physics Letters* , vol. 99, pp. 163102, October 2011.

[10] H. Dong, D. Z. Xu, J. F. Huang, and P. S. Change, "Coherent excitation transfer via the dark-state channel in a bionic system, " *Light: Science & Applications*, vol. 1, pp. e2, March 2012.

[11] S. Arijit, C. J. Lin, and K. C. Cheng, , "Single-Molecule Conductance through Chiral Gold Nanotubes," *The journal of physical chemistry*, vol. 117, pp. 13676, June 2013.

[12] H. H. Yang, Y. H. Chu, and C. I. Lu, "Digitized Charge Transfer Magnitude Determined by Metal-Organic Coordination Number," *Acs Nano* , vol. 7, pp. 2814-2819, March 2013.

[13] M. D. Croitoru, A. A. Shanenko, C. C. ,Kaun, and F. M. Peeters, "Ultra-small metallic grains: effect of statistical fluctuations of the chemical potential on superconducting correlations and vice versa, " *Journal of physics-condensed matter* , vol. 24, pp. 275701, Jule 2012.

[14] Y. H. Tang, V. M. K. Bagci, J. H. Chen, and K. C. Cheng, "Conductance of Stretching

Oligothiophene Single-Molecule Junctions: A First-Principles Study," *Journal of physical chemistry c*, vol. 115, pp. 25105-25108, November 2011.
[15] D. A. Luzhbin, and C. C. Kaun, "Bias-driven super-insulating state in prismane single-molecule contacts," *Physica Status Solidi-Rapid Research Letters* , vol. 5, pp. 265-267, June 2011.
[16] E. D. Kosten, J. H. Atwater, J. Parsons, and A. Polman, and H. A. Atwater, "Highly efficient GaAs solar cells by limiting light emission angle,"*Light-science & applications* , vol. 2, pp. e45, January 2013.
[17] X. Chen, B. Jia, Y. Zhang, and M. Gu, "Exceeding the limit of plasmonic light trapping in textured screen-printed solar cells using Al nanoparticles and wrinkle-like graphene sheets," *Light-Science & Applications* , vol. 2, pp. e92, August 2013.
[18] Z. C. Holman, S. D. Wolf, and C. Ballif, "Improving metal reflectors by suppressing surface plasmon polaritons: a priori calculation of the internal reflectance of a solar cell," *Light-Science & Applications* , vol. 2, pp. e106, October 2013.
[19] R. A. Taylor, T. Otanicar, and G. Rosengarten, "Nanofluid-based optical filter optimization for PV/T systems," *Light-Science & Applications* , vol. 1, pp. e34, October 2012.
[20] F. Pincella, K. Isozaki, K. and Miki, "A visible light-driven plasmonic photocatalyst,"*Light-Science & Applications* , vol. 3, pp. e133, July 2014.
[21] T. Hao, D. Kunpeng, K. C. Cheng, Q. Kuang, and S. Yang, "MoSe2 Nanosheets and Their Graphene Hybrids: Synthesis,Characterizations and Hydrogen Evolution Reaction Studies,"*Journal of materials chemistry a* 2, pp. 360-364, October 2014.
[22] G. S. Sahoo, and G. P. Mishra, "Design and modeling of an efficient metamorphic dual-junction InGaP/GaAs solar cell,"*Optical and quantum electronics* , vol. 48, pp. 420, June 2016.
[23] F. Azri, M. Labed, and A. F. Meftah, "Optical characterization of a-IGZO thin film for simulation of a-IGZO(n)/mu-Si(p) heterojunction solar cell," *Optical and Quantum Electronics s*, vol. 48, pp. 391, January 2016.
[24] E. Kadri, M. Krichen, and A. B. Arab, "Analytical method for the analysis of thin SiGe/Si solar cells with front surface field," *Optical and Quantum Electronics*, vol. 48, pp. 305, May 2016.
[25] F. L. Wu, H. Li, and Y. Tong, , "Tunable V-oc in polymer/ZnO array hybrid solar cells by ZnO interlayer with different nano-topographies," *Optical and Quantum Electronics* vol. 48, pp. 298, April 2016.
[26] Y. Kuang, D. Zhang, Y. Ma, Y. Liu, Z. Shao, X. Hong, X. Yang, and J. Feng,J, "Effect of near surface inverse doping on graphene silicon heterojunction solar cell," *Optical and Quantum Electronic* , vol. 48, pp. 199, February 2016.
[27] C. M. Chang, W. J. Ho, Y. T. Shen, S. K. Feng, and W. C. Liao, "Simulation of the optimized performance of thin-film silicon solar cells with nano-hole surface structures," *Optical and Quantum Electronic* , vol. 48, pp. 93, February 2016.
[28] D. M. A. Rahman, M. F. O. Hameed, and S. S. A. Obayya, "Light harvesting improvement of polymer solar cell through nanohole photoactive layer," *Optical and Quantum Electronic*, vol. 47, pp. 1443-1449, January 2015.
[29] Y. Sun, K. Li, J. Dong, X. Zeng, S. Yu, Y. Zhao, and H. Yang, "Effects of leakage current on the short circuit current in the dual-junction solar cells,"*Optical and Quantum Electronics* , vol. 47, pp. 433-440, April 2015.
[30] U. Aeberhard, "Effective microscopic theory of quantum dot superlattice solar cells, " Optical *and Quantum Electronics* , vol. 44, pp. 133-140, June 2012.
[31] A. Gagliardi, and A. Carlo, "Innovative structure for dye solar cells," *Optical and Quantum* Electronics , vol. 44, pp. 141-147, February 2012.
[32] S. Agnew, and P. Dargusch, "Effect of residential solar and storage on centralized electricity supply systems," *Nature Climate Change* , vol. 5, pp. 315-318, March 2015.
[33] X. P. Liu, X.H. Ding, Y. K. Ren, Y. Yang, Y. Ding, X. L. Liu, A. Alsaedi, T. Hayat, J. X. Yao, and S. Y. Dai,"A star-shaped carbazole-based hole-transporting material with triphenylamine

side arms for perovskite solar cells,"Journal of Materials Chemistry C , vol. 6, pp. 12912-12918, December 2018.

[34] S. B. Patel, A. H. Patel, and J. V. Gohel,"A novel and cost effective CZTS hole transport material applied in perovskite solar cells,"CrystEngComm, vol. 20, pp. 7677-7687, December 2018.

[35] X. Ma,Y. Liu, and H. Liu,"Fabrication of novel slurry containing graphene oxide-modified microencapsulated phase change material for direct absorption solar collector,"Solar Energy Materials & Solar Cells,vol 188, pp. 73-80, December 2018.

[36] J. Y. Xiang, X. C. Liu, M. N. Sun, X. G. Li,S. R. Wang, Y. Xiao, J. L. Zhang, and F. Zhang," Boosting the Stability of Perovskite Solar Cells through a Dopant-Free Tetraphenylbenzidine-Based Hole Transporting Material,"ChemistrySelect, vol 3, pp. 13032-13037, December 2018.

[37] S. Y. Hsun, F. K. Yuan, C. L. Shi,and Y.Y. Qian, "Surface plasmon resonance of layer-by-layer gold nanoparticles induced photoelectric current in environmentally-friendly plasmon-sensitized solar cell,"*Light: Science & Applications*,vol 1, pp. e14, June 2012.

[38] . Francesca, I. Katsuhiro, and M. K. Kazushi, " A visible light-driven plasmonic photocatalyst," *Light: Science & Applications* ,vol 3, pp. e133, January 2014.

[39] S. Y. Hsun, Y. H. Chia, C. C. Chung, L. T. Sheng, and S. H. Yun, " Ultra-thin titanium nanolayers for plasmon-assisted enhancement of bioluminescence of chloroplast in biological light emitting devices," *Applied Physics Letters* ,vol 103, pp. 063703, August 2013.

[40] Web of Science®

Effect of heat treatment on wear behaviour of rolled carbon steel in DOT4 brake fluid

Olawale O Ajibola[1,2*], Ojo J Akinribide[1], Samuel O Akinwamide[1], Peter A Olubambi[1]

[1]Center for Nanoengineering and Tribocorrosion, University of Johannesburg, South Africa

[2]Materials and Metallurgical Engineering Department, Federal University of Oye-Ekiti, Nigeria

Corresponding author: olawale.ajibola@fuoye.edu.ng

Abstract. The wear of heat-treated rolled carbon steels (HTRCSs) were performed under dry and wet sliding in DOT4 brake fluid is successfully investigated and reported in the study. The HTRCSs samples were obtained by heating to and soaking the as-rolled carbon steel (ARCS) at 950 °C prior to quenching or cooling in different media at different selected temperatures between (20 and 250 °C) and atmospheric conditions. The microstructures and wear track patterns of specimens were evaluated using SEM and high-resolution metallurgical microscope. The sliding wear behaviour and wear track patterns of the specimens impressed under 10N normal loads and reciprocating mode was used to assess the wear damages. The obtained wear properties were compared in air and DOT4 fluid environments using Anton Paar TRB tribometer. The results obtained showed that enhanced wear behaviours were obtained from the oil and water quenched samples as compared with the air and furnace cooled samples. The HTRCSs samples are more tolerable than as-rolled sample (ARCS) under dry sliding. However, beside the influences of the heat treatments, the lubricating effect of DOT4 is more pronounced in the results. The CoF ranges from 0.105 to 0.137 for wet sliding while higher CoF values (0.378-0.934) were obtained in the dry sliding.

1. Introduction

Apart many essential factors are considered by the automotive industry in a manufacturing of modern cars which include the high material strength-density ratio, reduced fuel consumption, safety and health enhancement and control of the dangerous exhaust gases [1]. Among various alloy, steel, cast iron and aluminium are widely used in making automobile automotive engine parts such as the engine blocks, cylinders and pistons of which the hydraulic brake cylinder is renowned. The high strength-weight ratio gives better advantage to aluminium alloy but very prone to wear problems and limitation to high temperature application. In contrast, cast iron and steel can withstand more aggressive application conditions. Rolled steel is famous in structural and industrial usage due to its combination of enhanced engineering properties controlled by several factors [2]. The heat treatment produces enhanced mechanical properties in steel alloys due to microstructural transformations [3-4]. Reinforced components can be made of multiphase structured steel alloys [5-6]. The displacement of the phase transformation proceeds with increasing Al, P and Si contents in the steel [7-8].

Until the recent time, reports were scarcely available on investigations involving wear and corrosion metal alloys such as (aluminium alloy) in mineral oils used for most hydraulic fluids (DOT3) and DOT4 fluids. Little explanations are made on wear theories of micromechanical surface contact [9] of aluminium alloy in hydraulic fluid. Thus, the current work is a fraction of the large study focusing on the alloys behaviour in hydraulic fluids (DOT3 and DOT4) of which some are available in literature [10-13]. In the present work, the heat treatment of carbon steel was done with the view to improving the hardness combined with toughness and strength for wear resistance hydraulic master brake application.

2. Experimental procedure

2.1 Material, Testing and Characterization

The 12 mm diameter low carbon steel specimen was received as rolled rod. Chemical analysis was done with Arc-Spectrometer (2000-3 Spetro-CJRO model). The chemical compositions of the bulk material and surface products were determined by Energy Dispersive X-Ray Spectrometer (EDS)facilities. The hardness test was perform using Innovatest - (Falcon 500 series) with dwell time of 10 sec on HV/0.2 Vicker's scale.

2.2 Heat treatment process

In all cases, the as-rolled specimens (12 mm diameter by 10 mm thick) were heated to 950 ^{o}C and homogenize at same temperature for 2Hr 30mins after which the samples were withdrawn from the furnace. H2 was vigorously quenched in oil bath (SAE-40 grade) pre-heated to and maintained at 250 ^{o}C constant temperature and left in the oil bath for about 30 mins before finally removing it. H3 sample was quenched in preheated water (105 ^{o}C) and removed from the water after 30 mins. The quenched H2 and H3 attained the baths temperatures within short times without the baths losing much heat. H4 was left in the open air to cool to the room temperature for 40 mins after the removal from the furnace whereas H5 was cooled in the furnace for 70 mins. The specimens were washed in benzene. Table 1 and Figure 1 present the conditions for the heat treatment and the cooling regimes for the experiment.

Table 1. Heat treatment cycles of rolled carbon steel.

Sampl e	Cooling media/	Cooling Temp. (^{o}C)	Cooling time (min)
H1	Control	-	-
H2	Oil	250	30
H3	Water	20	30
H4	Air	20	40
H5	Furnace	950~20	70

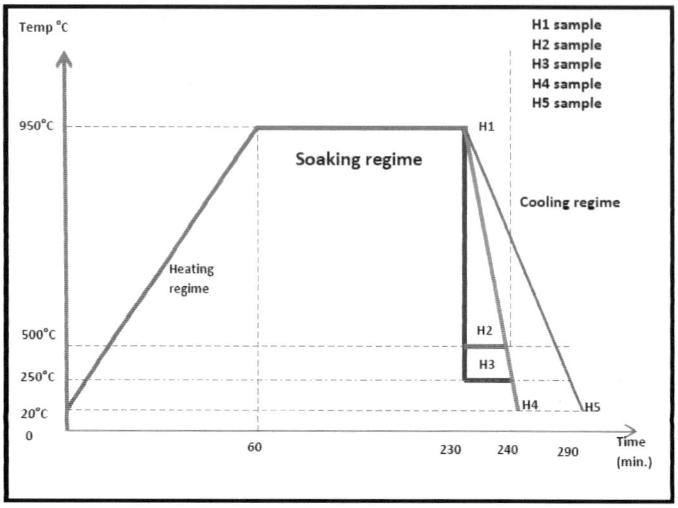

Figure 1. Heat treatment cycles used for the experiments.

2.3 Sample preparation and Metallographic examination

The samples were cut to 10 mm thickness using Brillant 220, ATM, Germany model automatic precision cutter and the surfaces were rough ground with P320 SiC grit paper and with Dia-max 9 µm poly suspension on Aka-Allegan 9 plate; polished with 3 µm poly suspension on Aka-Moran plate and with 0.2 µm fumed silica on Aka-Chemal plate on automatic polishing machine (Saphir 550, ATM, Germany). The surfaces were water washed for 3 sec and in methylated spirit to remove water on the surfaces before drying in warm fanned air. The samples were swab in Nital etchant according to ASTM-E407-70 [14]. The wear track patterns were examined by Scanning Electron Microscope (SEM-Vega3 Tescan model) and high-resolution metallurgical microscope.

2.4 Tribological tests of heat-treated samples

The wet frictional tests were performed on H1-H5 samples immersed in DOT4 fluid fed into the tribometer cup. The tribometer was set operating in the reciprocating mode using 10N normal load rubbed via 6 mm alumina ball against the test specimen surfaces. The obtained CoF values were plotted against the sliding time. The extent of wear damages and wear products were assessed based on the CoF values and visual examination of the wear track patterns observed on microscopes, SEM with EDS facilities. The obtained CoF (μ) values were compared under the dry and DOT 4 fluid environments.

3. Results and discussion

The chemical compositions and the micro-hardness of H1-H5 specimen are in Tables 2 and 3 respectively.

Table 2: Chemical compositions of as-rolled carbon steel

%C	%Mn	%Si	%P	%S
0.357	0.63	0.101	0.016	0.015
%Ni	%Cu	%Co	%W	%Fe
0.08	0.276	0.016	0.158	97.661

Table 3. Micro-hardness tests of rolled carbon steel

Samples	HV/0.2 (Average)
H1	168.74
H2	181.29
H3	289.88
H4	158.49
H5	115.68

3.1 CoF-sliding time patterns for H1-H5 under 10N load

The wear parameters of H1-H5 under the 10N reciprocating loading in dry and wet DOT4 fluid sliding are presented in Table 4. It relates the CoF with the track radius, wear cycle and sliding time.

Table 4. Wear parameters of steels under 10N load in dry and wet DOT4 sliding

Samples	Radius (mm)	Cycle	Slid time (s)	CoF	CoF max
$H1_{DOT4}$	3.99	14345	1434	0.131	0.227
$H2_{DOT4}$	11	5200	520.54	0.128	0.177
$H3_{DOT4}$	10	5712	571	0.105	0.216
$H4_{DOT4}$	7.63	7501	750	0.137	0.213
$H5_{DOT4}$	2.51	22780	2278	0.128	0.319
$H1_{Dry}$	7.39	5997	540	0.934	1.118
$H2_{Dry}$	11	4685	468.47	0.471	0.990
$H3_{Dry}$	10.5	4906	490	0.510	0.762
$H4_{Dry}$	4.62	12389	1238.98	0.378	0.533
$H5_{Dry}$	9.50	6028	542.43	0.741	1.118

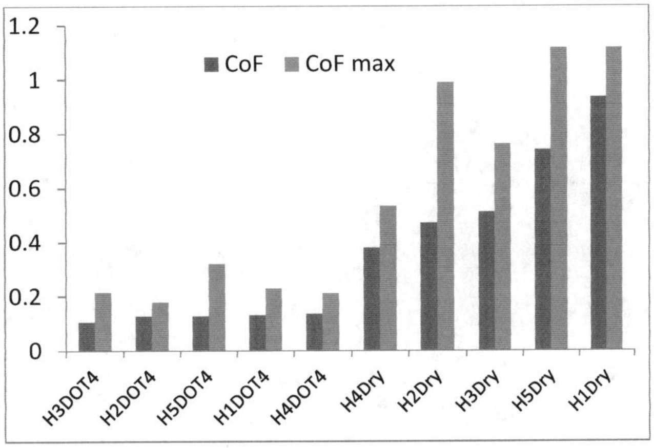

Figure 2. CoF variation in dry and wet DOT4 fluid sliding

Table 3 presents the micro-hardness tests of H1-H5 steel samples. The result demonstrates the effects of heat treatments on the hardness of the steel samples. The effects of oil and water quenching were more obvious on the higher micro hardness values obtained for H2 and H3 as compared with air and furnace cooled heat treated H4 and H5 samples. The H2 and H3 hardness increased due to quenching whereas the H4 and H5 samples hardness reduced due to softening. The wear parameters including the CoF of H1-H5 samples subjected to sliding wear under 10N loading in reciprocating mode of dry and wet sliding in DOT4 fluid are compared Table 4. It was observed that for all the samples (H1-H5), the CoF have better steady state in DOT4 fluid than the dry sliding wear. In Figure 2, the CoF and CoF max obtained from the dry and wet sliding in DOT4 fluid are compared for all the H1-H5 steel samples. In all cases, the dry wear test presents H1 as most affected. The mean CoF increased in the order of H1 > H5 > H3 > H2 > H4 for the dry sliding wear and H4 > H1 > H5 > H2 > H3 in the wet DOT4 fluid sliding wear condition. The effects of the heat treatments make H2-H5 samples to be more tolerable than H1 under dry sliding. However, beside the influences of the heat treatments, the lubricating effect of DOT4 is more pronounced in the results (Table 4 and Figure 2). The CoF ranges from 0.105 to 0.137 for wet sliding while higher CoF values (0.378-0.934) were obtained in the dry sliding.

3.2 Optical microscopy and SEM of microstructures of wear tracks

Fig. 3 shows the OM of microstructures of H1 to H5 test sample surfaces rubbed with alumina ball under the 10N load in dry and wet DOT 4 sliding contact environments. The effects of heat treatments are observed on Fig. 3. It illustrates and compares the severities of the wear in the dry sliding as shown in Fig. 3(a)-(e) and wet sliding in Fig. 3(a1)-(e1) conditions.

In Fig. 4(a)-(f1) the SEM and EDS of the microstructures and wear tracks of H1-H5 are presented. The EDS compares the compositions of wear products observed on the steel surfaces after dry sliding wear. Adhesive wear with particles formed at the tip of the asperities is observed and seem to be more pronounced in dry sliding whereas abrasion is significant in the wet sliding in DOT4 fluid [15-17].

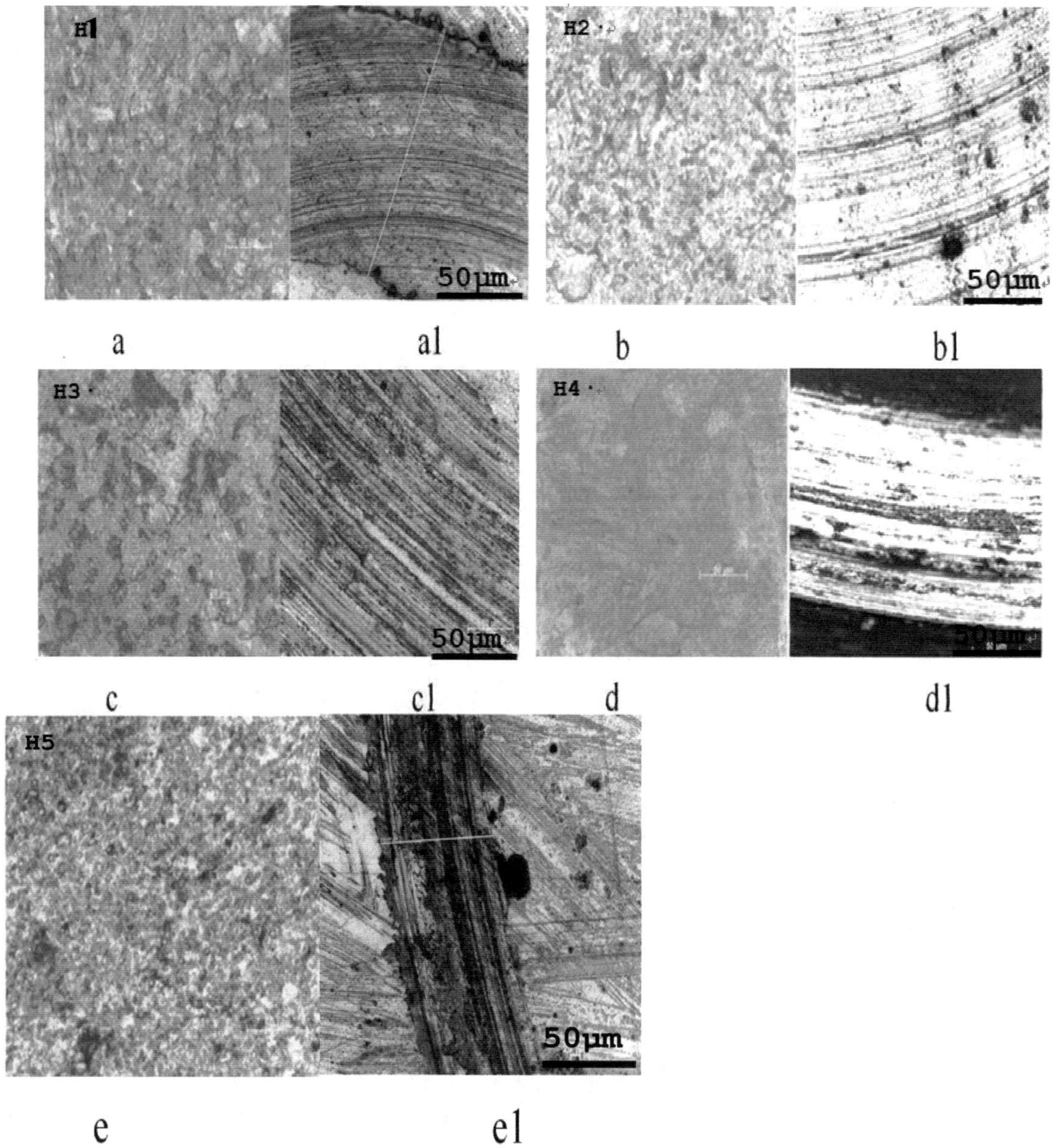

Figure 3. Optical micrographs showing the wear tracks of H1 (a, a1). H2 (b, b1). H3 (c, c1), H4 (d,d1) and H5 (e,e1) surfaces before and after wet sliding wear in DOT 4 fluid.

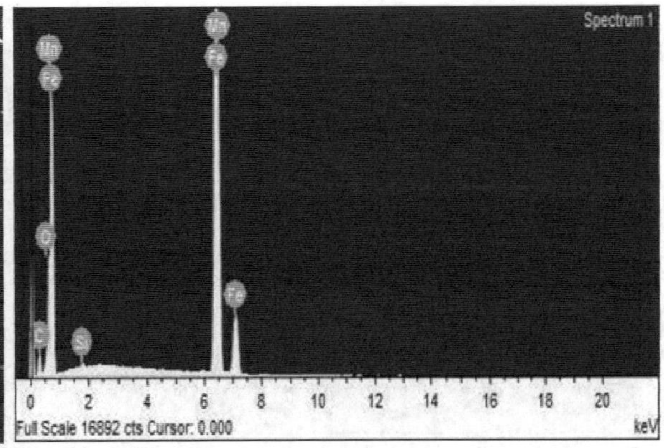
Spectrum 1
Mn
Fe
O
C
Si
0 2 4 6 8 10 12 14 16 18 20
Full Scale 16892 cts Cursor: 0.000
keV

a **a1**

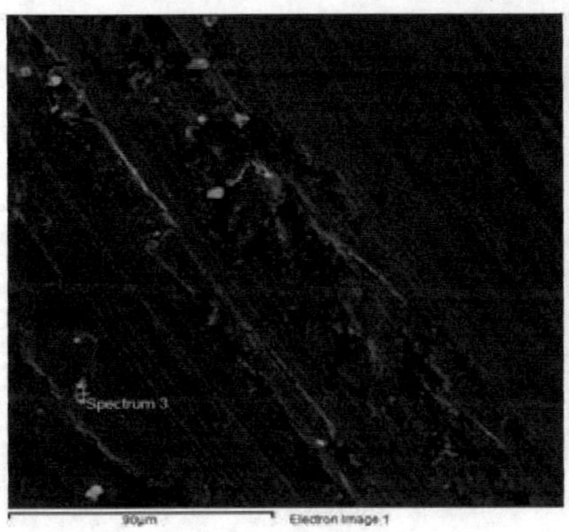
Spectrum 3
90µm
Electron Image 1

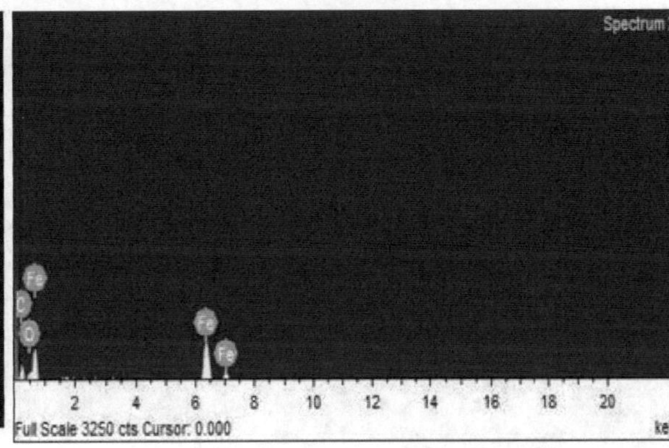
Spectrum 2
Fe
C
O
2 4 6 8 10 12 14 16 18 20
Full Scale 3250 cts Cursor: 0.000
keV

b **b1**

Spectrum 2
90µm
Electron Image 1

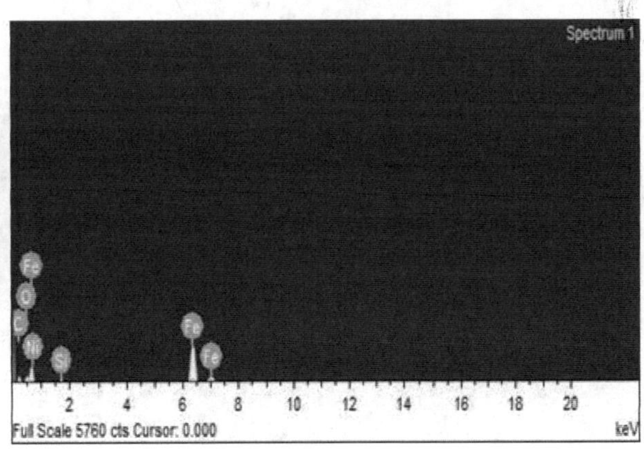
Spectrum 1
Fe
O
C
Ni
Si
2 4 6 8 10 12 14 16 18 20
Full Scale 5760 cts Cursor: 0.000
keV

c **c1**

IOP Conf. Series: Materials Science and Engineering **628** (2019) 012014 doi:10.1088/1757-899X/628/1/012014

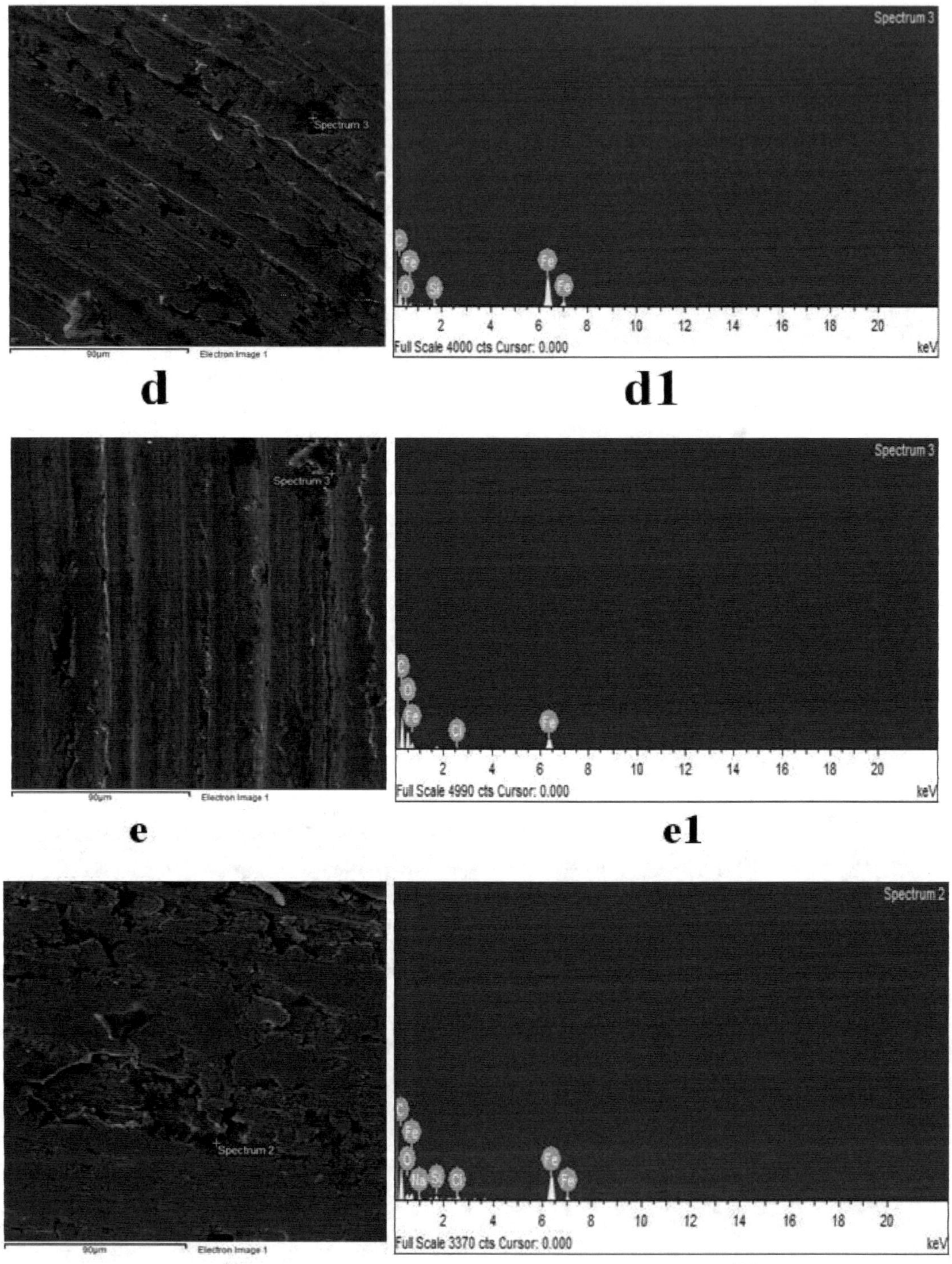

Figure 4. SEM/EDS of wear tracks of H1 (a, a1). H2 (b, b1). H3 (c, c1), H4 (d, d1) and H5 (e,e1) steels surfaces before and after dry sliding wear.

4. Conclusion

The following conclusions can be drawn from the results of the tests obtained; The heat treatments influenced the hardness of the steel samples. The oil and water quenching gave higher micro hardness values for H2 and H3 as compared with air and furnace cooled heat treated H4 and H5 samples. The H2 and H3 hardness increased due to quenching fine grained microstructures whereas the H4 and H5 samples hardness reduced due to softening. The CoF is in the order of H1 > H5 > H3 > H2 > H4 for the dry sliding wear and H4 > H1 > H5 > H2 > H3 for wet DOT4 fluid sliding. The heat treatments make H2-H5 to be more tolerable than H1 under dry sliding whereas the lubricating effect of DOT4 is more pronounced in the wet DOT4 fluid sliding results. The wear behaviour of H1 to H5 in Dot 4 brake fluid are largely depended on two wear mechanisms; adhesive wear and surface plastic deformation. The other important wear mechanisms such as spalling, and pitting were observed on dry samples. To conclude on the tribological behaviour of heat-treated steel in Dot 4 brake fluid environment, the selection of materials for this behaviour should mainly depend on the tribological sliding wear in dry and the lubricative wear. Surface fatigue resistance should be the major determinant for selecting a material for dry and lubricated wear.

5. References

[1] Adamczyk, J. and Grajcar, A., 2007. Heat treatment and mechanical properties of low-carbon steel with dual-phase microstructure. *Journal of Achievements in Materials and Manufacturing Engineering*,*22*(1), pp.13-20.

[2] Quest, C.F. and Washburn, T.S., 1940. Tensile Strength and Composition of Hot-rolled Plain Carbon Steels.*Trans. AIME*,*140*, pp.489-496.

[3] Bleck, W., 1996. Cold-rolled, high-strength sheet steels for auto applications. *JOM*, *48*(7), pp.26-30.

[4] Alderdice, R. and Campbell, E.A., 1998. Process plant implications of ULSAB and high strength autobody steels. *Ironmaking & steelmaking*, *25*(6), p.435.

[5] Bleck, W., Deng, Z., Papamantellos, K. and Gusek, C.O., 1998. A comparative study of the forming-limit diagram models for sheet steels. *Journal of Materials Processing Technology*, *83*(1-3), pp.223-230.

[6] Takechi, H., 2000. Application of IF based sheet steels in Japan. In *Proceedings of the International Conference on the Processing, Microstructure and Properties of IF Steels, Pittsburgh* (pp. 1-12)

[7] Adamczyk, J. and Grajcar, A., 2007. Heat treatment and mechanical properties of low-carbon steel with dual-phase microstructure. *Journal of Achievements in Materials and Manufacturing Engineering*, *22*(1), pp.13-20.

[8] Bera, B., 2013. Adhesive wear theory of micromechanical surface contact. *International Journal of Computational Engineering Resources*, *3*(3), pp.73-78.

[9] Ajibola, O.O. and Oloruntoba, D.T., 2015. Effect of MgFeSi inoculant on properties of Cast 6061 Al Alloy for brake master piston application. *Indian Journal of Materials Science*, *2015*.

[10] Ajibola, O.O., Oloruntoba, D.T. and Adewuyi, B.O., 2015. Effects of moulding sand permeability and pouring temperatures on properties of cast 6061 aluminium alloy. *International Journal of Metals*, *2015*.

[11] Ajibola, O.O., Oloruntoba, D.T. and Adewuyi, B.O., 2014. Effects of hard surface grinding and activation on electroless-nickel plating on cast aluminium alloy substrates. *Journal of Coatings*, *2014*.

[12] Ajibola, O.O., Oloruntoba, D.T. and Adewuyi, B.O., 2015. Effect of Processing Parameters on the Protective Quality of Electroless Nickel-Phosphorus on Cast Aluminium Alloy. *Journal of Metallurgy*, *2015*.

[13] Ajibola, O.O., Oloruntoba, D.T. and Adewuyi, B.O., 2015. Effect of Processing Parameters on the Protective Quality of Electroless Nickel-Phosphorus on Cast Aluminium Alloy. *Journal of Metallurgy*, *2015*.

[14] Ajibola, O.O., 2016. Evaluation of Electroless-Nickel Plated Polypropylene under Thermal Cycling and Mechanical Tests. *Tribology in Industry*, *38*(3).
[15] Ajibola, O., Komolafe, D. and Olorunfemi, B., 2016. Corrosion of NST60Mn and NST50 steels electroplated with copper in selected water environments. *FUOYE Journal of Engineering and Technology*, *1*(1).
[16] Ajibola, O.O., Ige, O.O. and Olubambi, P.A., 2018. Wear and Corrosion of Wrought A6061 Aluminium Alloy in DOT3 Brake Fluid. *International Journal of Engineering & Technology*, *7*(2), pp.512-519.
[17] Ajibola, O.O., Adewuyi, B.O. and Oloruntoba, D.T., 2014. Wear behaviour of sand cast eutectic Al-Si alloy in hydraulic brake fluid. *International Journal of Innovation and Applied Studies*, *6*(3), p.420.

Acknowledgement
The author would like to acknowledge the Management of Premier Wings Engineering Services, Ado Ekiti, Nigeria for providing the materials and some workshop services used for this work.

Institute of Physics
Dirac House, Temple Back
Bristol BS1 6BE UK

ISSN: 1757=8981
ISBN 978-1-5108-9832-5

9 781510 898325

CURRAN ASSOCIATES INC.
proceedings.com